VOID

Library of
Davidson College

From Protyle to Proton

William Prout (1785–1850). A portrait by H M Paget made in the 1880s from an original painting of the 1830s by John Hayes. The portrait was presented to the Royal College of Physicians in 1888 by Prout's youngest son, the Reverend Thomas Jones Prout. Courtesy the Royal College of Physicians.

From Protyle to Proton
William Prout and the Nature
of Matter, 1785–1985

W H Brock

Victorian Studies Centre,
University of Leicester

Adam Hilger Ltd, Bristol and Boston

© Adam Hilger Ltd 1985

All rights reserved. No part of this publication may be reproduced, stored in a retrieval system or transmitted in any form or by any means, electronic, mechanical, photocopying, recording or otherwise, without prior permission of the publisher.

British Library Cataloguing in Publication Data
Brock, W. H.
   From protyle to proton: William Prout and the nature of matter, 1785–1985.
   1. Prout, William   2. Matter
   I. Title
   539'.1'0924   QC171.2

   ISBN 0-85274-801-9

Consultant Editor: **Professor A J Meadows,** University of Leicester

Published by Adam Hilger Ltd
Techno House, Redcliffe Way, Bristol BS1 6NX, England
PO Box 230, Accord, MA 02018, USA

Typeset and printed in Great Britain at The Bath Press, Avon

Contents

Introduction		vii
1	William Prout (1785–1850)	1
2	Matter and Medicine	12
3	Physiology	34
4	Natural Theologian	60
5	Prout's Hypotheses	82
6	Prout's Molecular Theory and Biochemistry	109
7	The Prout Debate 1816–1860	143
8	Identical Outsides but Different Insides	179
9	The Bottomless Pit	217
Bibliography		226
Index		246

For Elvina

Introduction

According to the Aristotelian philosophical tradition, which lasted until the seventeenth century (and which William Prout absorbed as part of his reading in classics when a student at the University of Edinburgh at the beginning of the nineteenth century), the phenomena of material change studied by chemists were explained as follows. While gross perceptual alterations of chemical properties or substances were explained by the addition or subtraction of 'forms', an underlying substrate of matter—the *proto hyle* or primary matter—was believed to persist unaltered through the transformation. The forms or qualities of wet, dry, hot and cold informed or characterised the *proto hyle* to produce the four elements, Earth, Air, Fire and Water, which when mixed or compounded together, produced the definite material substances which were the primary concern of chemists. Although it might be possible to generate these elements from substances (for example, when wood is burned four elementary states appear to be produced) chemists in practice could usually only decompose a substance into its smallest homogeneous parts, the *minima naturalis*.

In the rival Greek atomic, or corpuscular, philosophy which had gained credence in the seventeenth century, chemical change was explained by the shapes, sizes, positions, motions and patterns of constituent microscopic particles which might, or might not, be defined as unsplittable corpuscles called atoms. Since each of these particles was homogeneous, it is possible to see a point of connection between the seventeenth-century corpuscular philosophy and the Aristotelian doctrine of primary matter. However, although enlivened by Boyle, Newton and their successors with gravitational force, chemical affinity and electrical properties, the earlier corpuscular philosophy or atomic theory was of little use to chemists until it was married to the modern doctrine of elements by John Dalton at the beginning of the nineteenth century.

It was only a generation before, in 1789, that the French chemist, Lavoisier, had finally eradicated the ancient four elements and defined elements operationally as substances not yet decomposed in chemical reactions. Such a pragmatic definition had enabled him to resolve the rather metaphysical notion of the four elements into some thirty or more tangible substances. Dalton's great contribution was to marry Lavoisier's elements with the old physical corpuscular, atomic tradition and to suggest that each of these distinctive elementary substances was composed from myriads of equally distinctive atoms. However, in assuming that there were as many different atomic particles as there were different elements Dalton abandoned at a stroke the age-old belief of philosophers in the simplicity of matter—that there was a unique, homogeneous primary matter. To many, perhaps most, British chemists and physicists who had been brought up in the rational tradition of natural theology, according to which God the creator had designed a perfect and simple world, it seemed inconceivable that God would have chosen to construct a universe from a large number of building blocks.

The scene was therefore set for William Prout to propose in 1816 that all of the so-called elements and their constituent 'atoms' were really compounds of one basic homogeneous material, which he called *protyle*. This he identified with hydrogen, the lightest known element. Prout's justification for this exciting, but nevertheless speculative assumption, came both from his classical training and from his attempt to reconcile information from Dalton's interest in the way substances combined together by weight with Gay-Lussac's evidence for the way gaseous substances combined by volume. Following some rather dubious assumptions and adjustments Prout was able to show that the relative atomic weights of all the known elements were whole numbers if the atomic weight of hydrogen was expressed as unity. From the integral nature of elementary weights it was then but a small step to infer that hydrogen was the primary matter from which all the 'elements' were composed. Thus, an element such as chlorine, with an atomic weight of 36, really consisted of 36 volumes of hydrogen compounded together.

These two hypotheses, of integral atomic weights and of the unity of matter, became known ambiguously ever after by Jöns Berzelius's term, 'Prout's hypothesis'. As a tantalising and attractive simplifying view of matter it was to be a continuous source of inspiration to chemists and physicists until the work of F W Aston on isotopes in the 1920s. On the one hand, for example, Prout's views received support from chemists such as Thomas Thomson, J B Dumas, J C Marignac and L Meyer; on the other hand, Berzelius, E Turner, J S Stas and D Mendeleeff opposed him. Yet whatever attitude individual experimentalists and theoreticians had towards the hypotheses, the work done in support or refutation proved incredibly fruitful. The hypotheses stimulated the improvement of analytical techniques, enforced interest in atomic weights and hence in the atomic theory of matter (which itself was treated with scepticism for much of the nineteenth century); they gave impetus to the search for a system of classification of the elements, so many of whose properties seemed similar, and, when the periodic law was revealed in 1869, they encouraged speculation about the evolution of the elements and structural theories of the atom.

Moreover, although with few exceptions physicists following Aston have failed to invoke Prout's name to justify their exotic explorations of the fauna of atomic particles (though in 1946 E E Witner proposed to name a unit of mass-energy the *prout*) it is clear to the historian that the conceptual search for a simple theory of matter remains as compelling as ever. Evidence that the nuclei of atoms were complex bodies consisting of protons and neutrons had been provided by Chadwick in 1932, and although a more complicated view than Prout's speculation that all material bodies were composed from hydrogen, the model of an atom composed from only three fundamental particles (protons, neutrons and electrons) still possessed the merit of simplicity and elegance. Unfortunately, even before the Second World War, the need to account for radioactive disintegrations and the presence in the atom of both weak and strong binding forces, had led both experimentalists, such as Anderson, and theoreticians such as Dirac and Yukawa, to predict and to discover new elementary particles such as positrons, neutrons, mesons and the extraordinary light-quantum. Despite these complications, physicists could take refuge in the simplicity offered by classifying particles according to whether they engaged in weak, strong or electromagnetic interactions. By 1960, however, following the intensive post-war study of cosmic rays and nuclear decay reactions in giant accelerators, the relentless growth in the number of elementary particles had dashed the dream of simplicity. Physicists found themselves in the same sort of position that chemists had been at the beginning of the nineteenth century when the number of chemical elements had increased embarrassingly.

Although natural theology was no longer invoked (or at least only indirectly in metaphors of God as the supreme mathematician) physicists instinctively reasoned that matter—as well as the forces which glued it together in a hierarchy of levels—must be simple. The principle of parsimony—the preference for theories which demanded few assumptions and the simplest explanations—therefore drove physicists, as it had chemists earlier, into a search for a Prout's hypothesis of the nucleus. The Proutian route taken by physicists in the 1960s (though not so recognised because, by and large, modern physicists are historically less well-informed than their forebears in the 1930s) invoked various principles of symmetry and algebraic group theories to produce classificatory simplicity. By sorting particles into a kind of periodic table, the 'eightfold way' implied the existence of another, more fundamental and mysterious, sub-stratum of matter, the quarks.

This Proutian theme of the roles of simplicity, aesthetics and patterning in understanding the nature of matter since the early 1800s forms the core of this book. But who was the man whose name is so memorably associated with this most dominant of themes in the history of matter theory? What else, if anything, did William Prout contribute to science?

In 1878, nearly thirty years after Prout's death, William Munk recorded in his useful biographical *Roll* of the Fellows of the Royal College of Physicians that he was unaware 'that any full and searching estimate of Dr Prout's merits as a philosopher and chemist has yet appeared'. Munk's remark remains true a century later and it is this gap which this book also tries to fill. Despite my use of as much surviving manuscript material as possible, Prout still remains a somewhat mysterious figure, overshadowed both by his older contemporaries like Dalton and Davy as well as by the next generation of chemists such as Faraday and Liebig. This has inevitably meant that the book is weighted towards intellectual biography and Prout the man, the physician, father and husband becomes absorbed into the discussion of Prout's contributions to matter theory, chemical analysis and the science which we now call biochemistry.

Between 1815 and 1827 Prout published a series of important papers on urine and digestion that opened up new areas of knowledge in purine chemistry and the chemistry of metabolism. In 1821 Prout published a concise clinical textbook on urine and the painful disease of bladder stones, but a similar work on the pathology of digestion, partly printed in 1822, was withdrawn from publication when his ideas were unsettled by his own brilliant, but controversial, demonstration that the gastric juices of animals and men contained hydrochloric acid. In 1840, however, he published a long and successful textbook of urinary and digestive pathology. Earlier, in 1827, he had classified foodstuffs into the still-accepted categories of carbohydrates (saccharinous substances), fats and proteins (albuminous substances) and published detailed analyses of sugars. As a teleological vitalist Prout believed that living systems (which were materially composed from organic compounds) also contained 'vital principles'. Under the direction of these principles, or directing agents, the three categories of *aliments*, together with water, were transformed into blood, a process he called 'primary assimilation'. 'Secondary assimilation' (which his younger German contemporary Liebig termed 'the metamorphosis of tissues') included both the processes of tissue formation from the blood and the destruction and removal of unwanted parts from the animal system. Such organisation of processed aliments could not occur, Prout was always to maintain, without the necessary presence of minute quantities of water or of elements other than carbon, hydrogen, oxygen and nitrogen.

All this proto-biochemistry may seem remote from Prout's hypothesis. As it turns out, however, apart from the intrinsic interest of his animal chemistry and its connections with the emergence of the discipline of biochemistry in the 1840s, Prout's biochemical speculations were based upon, and intimately connected with, his theory of the unity and simplicity of matter. There was, indeed, a unity about Prout's work, namely, a constant search for the laws which he felt sure governed 'not only the operations of the animal economy, but the whole material world'.

From Protyle to Proton falls into three sections. In the first, following a brief biography, there are two chapters concerned with Prout's contributions to experimental science by way of organic analysis, urine chemistry and the problems of digestion and sensation. Prout's contribution to the latter phenomenon also illustrates his strong speculative streak which is further analysed in the second section which consists of four chapters concerned with Prout's more theoretical work: a study of his *Bridgewater Treatise* of 1834 and of his support for natural theology and vitalism; and a detailed analysis of Prout's hypothesis and its reception until 1860, together with an account of his molecular theory and his attempt to link together his theoretical ideas with his experimental work in animal chemistry and human pathology. Finally, by way of an epilogue, and more impressionistically, the two concluding chapters of the third section examine the steps leading to a deeper understanding of the real basis for Prout's hypothesis in the 1920s through the emergence of the concepts of isotopes, protons and the structured atom; and the theme of simplicity which has continued to motivate particle physicists since the 1930s as they coined more and more elementary materials from the mysterious interiors of protons and neutrons.

Since I have worked on William Prout and his interests intermittently since the early 1960s my debts are too numerous to list exhaustively. They include a very large part of the British history of science community which has emerged during the same period, my colleagues at the University of Leicester, the many librarians and archivists who have helped me over the years, and my wife and my children who have grown up in the shadow of Prout. Particular debts of gratitude must, however, be paid to Prout's descendants, Lt-Col P E H Warner and Mr D Nicol for allowing me access to private papers, and to my typists Mrs J Nicol and Mrs A Gregson.

1

William Prout (1785–1850)

He was an example of a man gifted by nature with high intellectual endowments, improving these endowments by constant study, investigation and reflection. An amount of professional labour, such as would have wearied many men was daily performed by him; and from this he turned for relaxation to arduous chemical and mechanical researches. His mind was of that rare quality which is ever open to the reception of truth, and which steadily pursues that object, undismayed by difficulties, and indifferent to ridicule and neglect [Anon 1850:17].

The subject of these striking remarks, William Prout, was born on 15 January 1785 in the village of Horton, between Chipping Sodbury and Hawkesbury, Gloucestershire. The present village has altered little since the nineteenth century, and it still 'nestles easily in a fold of the hills which gird it round on three sides, but on the fourth slope steeply down, affording a lovely panoramic view of the fertile vale of Gloucester which stretches far away on either hand' (Hodges 1932). From the time of Edward VI until the 1780s the Lords of the Manor at Horton had been the great Catholic family of the Pastons but during Prout's youth the estate had passed into other hands.

In 1599 a certain Lawrence Allway purchased from the Pastons a parcel of land within Hawkesbury parish adjacent to the Horton parish boundary known as 'Chalkly' or 'Chalkleys'; and this land remained within the Allway family until the eighteenth century. To it was added, in 1652, the unentailed copyhold of 'Allway' in Horton parish, the lease being given by the Pastons who were then in debt from heavy recusancy fines. Finally, in 1723, a later descendent, William Allway, being without issue, left both 'Allways' and 'Chalkleys' to his nephew, Richard Bennet, whose sister Elizabeth was married to a Daniel Prout of Chipping Sodbury. This Daniel Prout was the great-grandfather of William Prout.

The Prouts, who may have originated from Flemish refugees, were a local family, for the surname is frequently found in various forms in the parish registers of Horton, Hawkesbury, the Sodburys and Wickwar, as well as in the *Horton Rental Book* of 1665–1741. Through their intermarriage with the Bennets they appear to have raised their station from that of peasant labourers and wool-workers to that of yeoman farmers.

In 1739, Richard Bennet's son, William Bennett (*sic*), purchased a freehold in Horton called 'Gilbert Ridings', and on his death, in 1777, he left the three homesteads of 'Chalkleys', 'Allways' and 'Gilbert Ridings' to his second cousin, John Prout (1745–1820), the father of William Prout. John Prout was the son of the splendidly-named Nebuchadnezzar Prout of 'Tungroves' (another copyhold of Horton) and Martha Hale; and a grandson of Elizabeth Bennet and Daniel Prout whose legal guardianship of William Bennett had led to this fortunate establishment of the family fortunes towards the end of the eighteenth century. However, because of the intricacies of the family tree, William Bennett's legacies did not flow straightforwardly to John Prout; moreover, since the lease on 'Allways' fell back to the Lord of the Manor in 1788, John Prout hardly inherited a fortune. Nevertheless, it must have sufficed to prevent him from emigrating to America to join two of his brothers who sent him fascinating reports of their attempts to make money from the real estate upon which the city of Washington was being built.

John Prout remained a Gloucestershire farmer, marrying a local girl, Hannah Lumbrick, in the 1780s. They had three children, the eldest being the physician and scientist, William Prout. The youngest child, Robert, died within two months of his birth in 1788, leaving the middle son John Prout, who was born in 1786, as the sole son to carry on the farming tradition at Horton until his death in 1862.

John Prout senior, who died grotesquely in 1820 'in consequence of having run a thorn into his hand which occasioned a locked jaw' (Anon 1820a), had in fact left the Horton estate to William. The latter, however, now a distinguished London physician, having no time or inclination for a country life, simply made over his rights to his bachelor brother who, in turn, willed them to William Prout's two eldest sons. The land, now reduced to a freehold of some 35 acres inconveniently dispersed within the parishes of Horton and Hawkesbury, remained in the possession of William Prout's great great grandsons until the 1960s.

Like so many other nineteenth-century physicians and scientists of humble origins, William Prout's early education was minimal. Although he learned to read and write at a dame school in Wickwar, and later attended a charity school at Badminton, his elementary education ceased before he reached the age of thirteen. From then, until he was seventeen, nothing is known of Prout's life, and it can only be assumed that he worked with his father on the land.

At the age of seventeen, however, Prout began to voice his dissatisfaction with his educational deficiencies, and with an awakened interest in mechanical things, mathematics and music, he determined to engage upon a systematic course of learning. With this aim, and presumably with his father's support, he left home for eighteen months between 1802 and 1804 to board with the Reverend John Turner (Anon 1848a), the Vicar of Horton and of the neighbouring Wiltshire parish of Luckington, at his Sherston Academy. Here he acquired the rudiments of Latin and Greek which were essential requirements for entry upon a university degree. However he returned home, either dissatisfied with his own progress or with the standards of Turner's Academy, and took the extraordinary step of advertising in a local newspaper for advice on the prospects for further education for an ill-educated twenty-year old man (Anon 1851).

A reply came from another clergyman, the Reverend Thomas Jones (1758–1812), who ran a Classical Seminary at Redland, Bristol. Jones had been educated at Cambridge and Dublin, and had been Vicar of two Devonshire parishes before he opened his 'classical seminary for young gentlemen' at the turn of the century. Jones was later described as a liberal and generous man, and in view of Prout's obvious admiration for him, hallowed by the christening of his youngest son with the names Thomas Jones, it may be significant that:

> Dr. Jones had himself experienced the difficulty of emerging from obscurity and comparative indigence, to distinction and competence: and to those who had engaged in the same arduous struggle, his advice and assistance were always accessible. Many were indebted to him for the first step in the progress of their advancement, and therefore we may hope that many will love his memory, as long as recollection shall hold its seat in their bosoms [Evans 1816:ii, 350].

Prout spent two happy and formative years with Jones. In return for his tuition he taught the younger pupils of the school, and stimulated by a pupil's curiosity over the exciting new events that were then taking place in electrochemistry in the hands of Humphry Davy and others, Prout began to form what was to become a lifelong passion for the science of chemistry. It was Jones who suggested that Prout should become a doctor and who recommended him to enter the medical school of the University of Edinburgh. Thus, in 1808, at the mature age of twenty three, Prout went up to Edinburgh, armed with a letter of introduction to Jones's old teaching friend, the famous grammarian Dr Alexander Adam (1741–1808), Rector of Edinburgh

High School. He was to remain diligently in Edinburgh for three years except for visits with the Adam family to the Highlands and to the Edinburgh suburban villages of Duddingston and Morningside during the long summer vacations. These villages, we are told by an obituarist, were sufficiently close to the University to ensure full use of the library during the summer. Alexander Adam's death during Prout's second year in the city seems to have brought him close to Adam's widow and her two children, Walter (1792–1857) and Agnes (1793–1863). Walter soon became Prout's confidant, while by the time Prout graduated in 1811 there was an implicit understanding within the family that he would marry Agnes as soon as he had established himself in the world.

Edinburgh University was then at the zenith of its fame. Its reputation

> had been raised to an unparalleled height by the Monros, by Black, by Cullen, by the Gregorys, the Homes, the Rutherfords, the Hamiltons, the Bells; by Barclay, Gordon and many others; whilst the general fame of the University had been sustained by such names as those of Robertson, Blair, Hutton, Dugold Stewart, Playfair, Leslie, Brown, and a phalanx of eminent theologians, philosophers and literati [Hall 1861; Morrell 1971].

Its medical graduates, in the long term, helped to create the general practitioner and to break down the traditional hierarchy of physician, surgeon and apothecary; moreover, its graduates, being well-versed in chemistry, botany and physiology, showed a tendency to contribute as much to these pre-clinical sciences as they did to the art of healing.

Unfortunately, very little is known of Prout's days as a medical student for, like the vast majority of undergraduates, he left little mark upon his university or on the many medical societies of Edinburgh, though an elaborate paper on *The history of physic*, which he read to an unknown audience on 10 August 1809, has survived, as have Latin essays on hearing and sensation. 'You know how little I mix with the world', he confided to Walter Adam, 'and how limited [*sic*] my acquaintance is' (Prout 1810b). We know that his teachers included Alexander Monro *tertius* for anatomy and surgery, Thomas Hope on chemistry and pharmacy, Andrew Duncan Jr for physiology and the history of medicine, and James Hamilton for midwifery. In addition, Prout would have attended clinical lectures on the patients at the Royal Infirmary. Although his fellow students included Marshall Hall, who was to investigate the nature of reflex action, John Davy, the brother of the chemist Humphry Davy, and Henry Holland, a future Royal physician, his only close friend appears to have been John Elliotson. Elliotson was a highly original physician whose reputation was to be clouded in the 1830s by his use of mesmerism and his championship of phrenology. However, Prout's references to the extramural chemistry teacher, Thomas Thomson (Morrell 1968–9) in his letters to Walter Adam, imply more than just social acquaintance. Thomson's decision to abandon Edinburgh in 1810 in order to seek fame and fortune in London seems to have been watched with interest by Prout.

Although he had already completed a tedious, and not very original, Latin dissertation on fevers, their definitions, symptoms, pathology, prognosis and treatment by August 1810, Prout did not qualify as MD until 24 June 1811. One of his teachers, the anatomist Mathew Baillie, advised him against trying to practise medicine in Edinburgh or elsewhere in Scotland and, with his future unclear, Prout returned to Horton. From here he went to help his former teacher, Thomas Jones, put his farm lands at Westbury in Somerset in order. Inquiries in the West Country soon showed that establishing a practice in wealthy Bath or Bristol would be an uphill task. Although a country practice at Wells fell vacant Prout decided not to take it because the extensive travelling it involved would have prevented him from having time for the anatomical and physiological investigations he wanted to make. 'My health', he told Mrs Adam, 'is better than I ever knew it—so much exercise and particularly riding, have quite renovated me after my writer's studies. I am become so fat you would scarcely believe me to be the same person' (Prout 1811b). Restored by his Somerset ramblings, Prout moved to London in October 1811 to investigate the possibility of becoming a lecturer in a private medical school or in one of the metropolitan hospitals. More particularly, he needed to obtain a licence to practise medicine from the Royal College of Physicians. Accordingly, and politely refusing financial help from Mrs Adam (Prout 1812b), he took rooms in Leicester Square where he could conduct nocturnal chemical experiments on the injection of anatomical specimens for teaching purposes (Prout 1813c). During the day he walked the wards of the united hospitals of St Thomas's and Guy's, before presenting himself for the College's oral Latin examination on 22 December 1812. This examination, which cost the large sum of £50, consisted of routine questions on physiology, pathology, therapeutics and the interpretation of a passage from either the Roman authority, Celsus, or the seventeenth-century physician, Thomas Sydenham. It clearly presented Prout with no difficulty for he dismissed it to Walter Adam as a 'trifling' experience.

'Ambition prompts me to look to the summit of my profession and "no middle course to steer"', he told Adam. Now a Licentiate of the Royal College of Physicians, and having already published two short papers in the *London Medical and Physical Journal*, Prout set up his practice at 4 Arundel Street, just off the Strand, to bring his name forward within the metropolitan medical community. This was his most dismal period in which, as we learn from his frank outpourings to Walter Adam, he thought seriously of abandoning medicine altogether. On 19 September 1813 he confessed:

> The chance of success daily gets less with me. I expect to be obliged to give up all thoughts of lecturing [in a teaching hospital], if so, I do seriously think, between ourselves, that I shall for ever quit the profession and return to some spot where I am totally unknown. Perhaps leave the country for ever [Prout 1813b].

And to Mrs Adam, Prout (1814b) revealed something of the social difficulties that someone like himself faced in attracting patronage and added:

> The state of society here is very wretched in my opinion and entirely artificial. Would you think it possible that one can meet twenty or thirty human beings without hearing a rational sentence? Eating and drinking, the superior merits of a favourite snuff, the abilities or want of abilities of some poor actor or actress, constitute the whole topics and are subjects so warmly discussed as if the salvation of the whole human race depended upon them. What vexes me most is, that our females, who I am willing to believe are gifted by nature as bountifully with common sense as they are elsewhere, are thought to despise it, and reduce their understandings to the despicable level of those of coxcombs and fools. This degradation of female intellect, is what we call *genteel education*, and a miss is said to be the most *genteely educated* whose understanding is the most degraded or obliterated, or who has the least share of common sense.

In order to announce himself as a man of science, and to seek the patronage that would catalyse his search for medical clients, Prout delivered a course of lectures on animal chemistry at his home between February and April 1814, 'the attendance on which though small was select, and so highly was he already esteemed that his audience constantly included [Sir] Astley Cooper' (Munk 1878: iii, 109). Prout's prior attendance at St Thomas's and Guy's hospitals would have brought him into contact with this great surgeon who, in the early 1800s, had belonged to a number of home-based metropolitan clubs and societies which were devoted to chemical and physical investigations. Another of Prout's auditors, who became a close friend, was Alexander Marcet, a pioneer animal chemist and correspondent of the Swedish chemist, Jöns Berzelius. From a remark in Berzelius's travel journal of his second visit to London in 1818, it is clear that Prout's ambition was early election to the Royal Society, since this would give him an *entré* into the aristocratic world of its President, Sir Joseph Banks, and oil the wheels of his growing medical practice. In the event, Prout did not become a Fellow of the Royal Society until a year after Berzelius's visit; moreover, Berzelius also recorded Prout's unhappiness at not being given membership of the socially exclusive Chemical Club, to which Cooper, Marcet and other London hospital physicians belonged.

Two advertisements for Prout's lectures were published in the *Annals of Philosophy* (1813; 1814), the important monthly periodical which Thomas Thomson had begun in 1812 and to which Prout was to submit much of his research.

> Dr. Prout intends in the course of the winter, to deliver a series of lectures on Animal Chemistry. The object of these will be to give a connected view of all the principal facts belonging to this department of chemistry, and to apply them, as far as the present state of our knowledge will permit, to the explanation of the phenomena of organic action.

> Dr. Prout will commence a series of lectures on Animal Chemistry on Friday, February 18 [1814], at half past eight in the evening. These lectures will be given at his residence, 4 Arundel Street, Strand, and will be continued weekly at the same hour.

A heap of Prout's lecture notes has survived. Although almost indecipherable, it is sufficiently clear that he first offered his audience a general statement of his chemical philosophy—to be discussed in the following chapters—followed by a detailed review of the chemistry of the blood, respiration, digestion and the urine. Some of this material formed the basis of his only extensive paper, that on sanguification, first published in 1816 and republished in 1818. From references in his later publications it seems highly probable that he first revealed his method for the preparation of pure urea at one of his lectures, as well as the details of some of the organic analyses that he had begun to make.

If his lectures did not bring him election to more socially exclusive clubs like the Chemical Club and the Royal Society, Cooper and Marcet were able to ensure his election to the Medical and Chirurgical Society on 10 May 1814. Although his connection with this prominent medical society—'the most respectable in the city' (as he termed it to Adam)—lapsed after 1835, he read several papers to it, served on its Council from 1817 to 1819 and was one of its Vice-Presidents in 1823 and again from 1833 to 1835. That this election occurred only just in time to cure Prout's acute sense of failure is clear from a letter he wrote to Adam only three weeks before:

> I have serious thoughts of going to France this summer perhaps to reside there or in Germany. My chance of success here is very little or nothing and fear will render it still less ... As for staying in London in the way I am now going on, even for another winter, it is impossible. I cannot do it, and I am determined not to make the attempt [Prout 1814c].

However, stay he did and confident at last of his ability to succeed as a physician, Prout was married on 22 September 1814 to Agnes Adam. Europe being momentarily at peace, the married couple were able to pay a visit to Paris from their honeymoon base in Ramsgate. In Paris, like Humphry Davy and his bride, they were able to have a private view of the artistic treasures which Napoleon had looted during his campaigns with the aid of Baron Denon. A *Medical Times* obituarist rhapsodises over this episode: 'We can imagine the delight with which a man so refined, and so open to the appreciation of the beautiful as ... William Prout—no mean draughtsman himself—wandered through the noble salons of the Louvre'. One of Prout's own watercolours—a painting of Horton Church in 1804—has survived and does indicate a delicate artistic imagination. Both he and his wife evidently enjoyed having paintings about them, for the same unknown writer reports that Prout's consulting rooms were hung with many canvases. Later in life, Prout became a friend of the artist Thomas Phillips, whose son was to make the posthumous copy of the portrait of Prout made by John Hayes.

On their return to England the Prouts settled in a new home in Southampton Street, Bloomsbury, where a daughter was born in 1815. The child survived only a few months, but there were six further healthy children: John William, born in 1817, who became a successful lawyer; Alexander Adam, born in 1818, who became an army doctor; Walter Robert, born in 1820, who lost his life as a Major during the Indian Mutiny; Thomas Jones, born in 1823, who took holy orders and became a reforming classics don at Christ Church, Oxford, and two daughters, Elizabeth, born in 1825, and Agnes, born in the following year.

Prout's attempts to become a Fellow of the Royal Society were finally rewarded in March 1819, His proposers, who were all doctors, included his friend, Alexander Marcet, the polymath, William Hyde Wollaston, and a group of distinguished medical men, Henry Warburton, Charles Koenig, Peter Mark Roget, Leigh Thomas, Gilbert Blane and Mathew Baillie. He was to serve the Society well by acting on its Council from 1826 to 1828 (he declined the service in 1832) and by supervising the construction of the Society's standard barometer in 1835 (Brock 1970). In return, the Society honoured him with its highest accolade, the Copley Medal, in 1827. Two years later, he was made a Fellow of the Royal College of Physicians and in 1831, as one of the four youngest Fellows of the College, he was invited to deliver the Gulstonian Lectures. His three lectures on 'the application of chemistry to physiology, pathology and practice' were summarised in the *Medical Gazette*, where they gave rise to an acrimonious dispute with another College Fellow, Alexander Wilson Philip. Personal abuse and invective were still common weapons in controversy, but this debate was particularly sterile and tedious. Annoyed by Prout's remarks on physiologists' lack of chemical knowledge and insight, Philip catalogued his own chemical and physiological insights in great detail. A diplomatic, but non-committal, reply from Prout, was followed by the accusation that Prout was rude and careless with words; Prout replied hotly that Philip was over-fond of wordy disputes, fault-finding and expounding the 'same thrice-told tale' of his own achievements. And so the dreary dispute rolled on until Prout finally had the sense to withdraw from the argument and allow Philip the egotistical last word (McMenemey 1958).

These few facts are virtually all that is known of Prout's life; for although Agnes survived her husband by thirteen years, unlike many Victorian widows of distinguished men, she did not publish any loving study of her husband's achievements as 'the first chemist of the metropolis' (Murchison 1831). Only rarely, too, did a scientific contemporary mention Prout's name, and then never to reveal any personal facet of his life. Since virtually none of Prout's personal correspondence has survived, the remainder of his life after 1814 has to be summarised in the generalisation that he became a very successful physician who specialised in urinary and digestive complaints from his Sackville Street home in Piccadilly where the family had moved in 1825; the details of his life after 1814 until his death then become those of his scientific books and papers, and of his persistent scientific influence not only on matter theory, but also in biochemistry.

The intense earache, which obituarists mention he suffered as a youth, seems to have been a symptom of the deafness that afflicted him in early middle age, causing him to shun scientific society. Thus, there is no mention of him participating in the affairs of the Royal College of Physicians after the Gulstonian Lectures in 1831, or at meetings of the Royal Society after 1841, or at the British Association for the Advancement of Science after 1842. Significantly, he declined an invitation to be a founder member of the Chemical Society in 1841. Deafness must have been especially tragic for him musically. At some stage in his life he built an organ 'which he played with great skill; several anthems were also composed by him' (Anon 1850).

Soon after the publication of the final edition of his textbook on urinary diseases in 1848, he fell a victim to the cholera outbreak of that year. Although he recovered, he collapsed again during the summer of 1849. An autumn excursion to the Chalkley estate in Gloucestershire did not improve matters. Emaciated, he returned to London to continue with his practice. His health grew steadily worse in the spring of 1850, and when Sir Benjamin Brodie, the surgeon President of the Royal Society, called to see him on 9 April, Prout told Brodie that he was dying. The end came the same day, the cause of death being attributed to gangrene of the lung; Prout, like Humphry Davy, but curiously for a doctor, had requested no post-mortem. A notebook bearing some analyses of sugars he was making was found in his pocket and carefully preserved by Agnes. He was buried in Kensal Green cemetery, his sons placing a simple memorial tablet in Horton church.

> Sacred to the Memory of William Prout, M.D., F.R.S.
> Born in this parish 15 January 1785
> Died in London 9 April 1850
> *Scintillulam contulit*

Prout was an early riser who, like Lavoisier, did most of his research and writing before he breakfasted at 7.00AM, the remainder of the day being devoted to his patients. 'Besides his extensive town practice', we are told, 'scarcely a day passed that boxes and parcels did not arrive from the country, and the analysis of their contents, together with the necessary correspondence, consumed no small portion of each day' (Anon 1850; Anon 1851). This implies, perhaps, that provincial doctors regularly consulted Prout by sending him pathological samples of urine and urinary stones for analysis and seeking his advice on diagnosis and an appropriate regimen and treatment.

> In stature he was a slim man of medium height. His head was nobly developed, and his intellectual qualities strongly marked; the hair soft and snowy white. His features were delicately chiselled, eyes brilliant, complexion very pale, but the expression of his countenance combined benevolence with great intelligence. There was a blandness in his manner which inspired confidence, and set the most nervous patient at ease. He always dressed with scrupulous neatness, usually in black, with gaiters, or silk stockings [Anon 1850].

This phrenologically-inspired description aptly fits the surviving portraits of Prout by John Hayes and the copies made by H W Phillips and H M Paget for the Royal College of Physicians and the

University of Edinburgh. To *The Lancet* (1850) obituarist he was 'the *beau ideal* of the venerable physician'.

Despite his deference and obscurity, Prout impressed those with whom he came into contact. Thomas Thomson was highly struck by him, while William Charles Henry (1854), who disagreed profoundly with Prout's interpretation of the atomic theory, admitted that, like Wollaston and Davy, Prout possessed 'a taste for extreme exactitude' and an 'unrivalled manual expertness' never achieved by John Dalton. Berzelius, on meeting him for the only time in August 1818 had 'expected to find a quite young man and from the manner of his writings a very resolute gentleman'. Instead, to his surprise he met

> a middle-aged man, fully grey-haired and with a grey beard which was, contrary to what is usual in England, somewhat long. His entire disposition was one of modesty and in his gestures [*åtbörder*] there was a certain degree of outspokenness [*rättframhet*], if I may use that word, from which all the edges and corners had been removed by the lathe (an operation which I rather wish had at some time occurred to me) [Berzelius 1903: 112].

The Lancet (1850) obituarist, already quoted, captured all these feelings in three telling sentences:

> The influence of his genius will be felt to the remotest generation, wherever chemistry, in its varied and important applications to physiology, pathology and therapeutics, to health, disease, and remedies, shall be studied. It is an honour to our profession and country to have produced such a man; it should be felt as a universal regret that his career of usefulness and investigation has closed. He has died, however, full of years as of honours, and 'the silver chord' was not 'loosed' until he had raised in his works an imperishable monument of his fame.

2

Matter and Medicine

> I am engaged in the investigation of the laws which [molecular weights] obey—laws which appear to regulate not only the operations of the animal economy, but the whole material world (Prout 1818a: 484)

In a letter published in 1822, Prout wrote that he had been inspired to enter into the field of organic chemistry by the hope that a discovery of the laws which governed the combination of carbon, hydrogen and nitrogen in organic compounds would lead to 'an insight into the laws which regulate the union of other elementary principles'. To this end he had

> set to work, and after a very great labour, and no trifling expense in apparatus, etc., succeeded, as I supposed, in analysing more or less perfectly almost every well-defined and crystallized organic substance that I could procure. A few of my earlier results were published, perhaps prematurely, but the great mass, as is well known to several of my friends, still remain by me, nor have I for various reasons, the least inclinations to publish them at present [Prout 1822a: 424].

Since the analyses which he released for publication in 1827 were relatively few in number, it seems that little of this 'great mass' was subsequently published. Prout was clearly a perfectionist who was unprepared to publish authoritative analyses until he was certain in his own mind that his technique was accurate and hardly open to improvement. However, this decision was, by the 1830s, out of keeping with the practices of 'gentlemen of science'. In the end, it lessened his fame and robbed him of the essential stimulus which comes from mutual scientific criticism.

Although a belief that chemical phenomena were the basis of physiological processes was quite common among early nineteenth-century chemists, especially among those interested in vegetable and animal chemistry, it was not always so recognised by physiologists (Mendelsohn 1964; Holmes 1965; Holmes 1974). Part of Prout's aim, and achievement, was to alter this state of affairs; what he began Liebig and Bernard completed. However, chemists were not able to speculate about the details of physiological chemistry until they first had a knowledge of elementary compositions. This knowledge, moreover, had to be both accurate and extensive. The fact that Prout spent much time in the search for accurate methods and that he published relatively few analyses compared with Liebig (who in any case had the advantage of a constant flow of pupils to act as his assistants), did not, however, prevent him from applying his knowledge to physiology and medicine.

The Art of Analysis

The analysis of chemical substances has been advanced by chemists at two levels: by the resolution of substances into either their 'proximate' (immediate) principles (e.g. the decomposition of a carbonate into an oxide and carbon dioxide), or into their ultimate elements (elementary analysis) (e.g. the analysis of lime into the elements calcium, carbon and oxygen). Although both these kinds of analysis may be pursued qualitatively or quantitatively, historically the analysis into proximate principles was associated with qualitative ends, whilst elementary analysis, which usually succeeds proximate analysis, was associated with quantitative procedures. In the divisions of chemical science known as vegetable and animal chemistry (roughly equivalent to today's organic chemistry), the techniques of proximate analysis were developed from the seventeenth century onwards and systematised by the post-Lavoisier French chemists Fourcroy (1801–2), Vauquelin and Chevreul (1824). However, with the full acceptance of Lavoisier's oxygen-based antiphlogistic chemistry and the appearance of Dalton's atomic theory in the early 1800s, the methods of elementary analysis pioneered by Lavoisier himself, Gay-Lussac, Thenard, Prout and Liebig, proved themselves more important for the development of organic chemistry and biochemistry.

Although Prout used both kinds of analysis, he was particularly associated with the quantitative determination of elementary composition. His development as an analyst took place during the decade between 1815 and 1825, and by the end of this period he had established a European reputation as an accurate and authoritative experimentalist. Hence, before looking at his ideas and findings in the specific fields of the chemistry of urine and of digestion, it is appropriate to say something concerning his actual contribution to the advancement of gravimetric analytical apparatus for the purpose of elementary organic analysis.

Quantitative organic analysis began effectively in 1784 when Lavoisier analysed alcohol, wax, olive oil and sugar by direct combustion of these substances in oxygen. The carbon dioxide formed in the combustion was absorbed in caustic potash (from which the carbon content could be determined) and the water formed was calculated by weight differences. However, this method was cumbersome, inaccurate and sometimes dangerous, and in any case, Lavoisier's execution in 1794 prevented him from refining the method. It was Gay-Lussac and his collaborator Thenard who in 1810 first introduced a technique which was capable of producing reliable and, therefore, significant results. In their method, the organic substance to be analysed was mixed into a paste with the oxidising agent, potassium chlorate, and heated in pellet form in a specially constructed apparatus. Although the technique was time-consuming, dangerous and dependent upon the dexterity of the operator for its accuracy (Berzelius 1814b: 402) the French chemists were able to apply it with skill and reasonable accuracy to the analysis of a variety of animal and vegetable substances, including albumen, sugars and several vegetable acids. One of their more interesting theoretical conclusions was that the proportions of hydrogen and oxygen in sugars and mucilaginous substances (starches, gums) were similar to the proportions of these elements in water. Such substances, they suggested, might be hydrates of carbon (Gay-Lussac and Thenard 1811: ii, 321–3). Prout was to adopt this suggestion in the family of organic substances he called 'saccharinous', the more familiar term 'carbohydrate' not being introduced by Carl Schmidt until 1844.

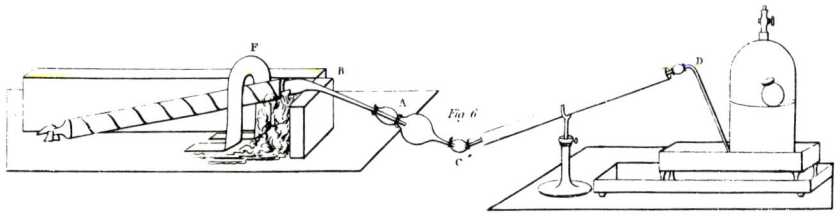

Berzelius's potassium chlorate combustion apparatus for the elementary analysis of organic compounds (from *Ann. Phil.* **4** (1814) Plate XXV).

After 1814, organic analysts (including Prout) adopted Berzelius's simple horizontal arrangement of the French chlorate method (Berzelius 1814b). This had the advantage that elaborate calculations were avoided by estimating carbon and hydrogen directly from the absorption of carbon dioxide and water in caustic potash and calcium chloride. Moreover, the danger of explosions, which were only too frequent with the French apparatus, was avoided by controlling the rate of combustion over a charcoal fire by a movable iron screen. These features, with modifications, were adopted by Prout in his first analytical publication in 1815. Where he most differed from Berzelius was in his advice that it was better to make separate combustions for the estimation of individual elements, than to attempt to estimate the quantities of the constituent elements from one combustion as Berzelius did. This would, of course, have lengthened the analytical proceedings. Moreover, he thought (incorrectly) that Berzelius had overlooked the fact that any water vapour produced during the combustion could interfere with the weights of the other gaseous products and hence produce a hydrogen estimation above the true one; this is perhaps the origin of Berzelius's exasperation with Prout in letters to Marcet—an exasperation which evaporated, however, once they met in London in 1818. It was for this reason though that Prout developed a drying apparatus for the exsiccation of substances before analysis.

In the gravimetric analysis of organic compounds Prout (1815a: 273) demonstrated how W H Wollaston's equivalent slide-rule, or 'synoptical scale of equivalents', could be modified so as to produce the empirical formula of a substance, assuming (as Berzelius had declared after some hesitation in 1814) that organic substances obeyed the law of definite proportions. Providing the proportions of at least two of the elements were known, then the number of atoms of each element in a ternary or quaternary compound was easily discovered. If the scale of a manufactured Wollaston slide-rule was pasted over with a new logarithmic organic scale bearing only 'multiples of an atom of oxygen, hydrogen and carbon, from one to ten; and azote [nitrogen], from one to four or five or more', the instrument could be adapted for organic estimations. For example:

> suppose we had the weight of a particle of a ternary compound to be 46.5, oxygen being 10, and that 46.5 parts of it contained 15.15 carbon, 1.34 hydrogen, and consequently 30.01 oxygen. To find the number of atoms of each of these elements, we have only to place 10 on the slide opposite oxygen, and then opposite each of the numbers respectively we have the numbers of atoms of each element required. Thus opposite 15.15 carbon we have 2 carbon; opposite 1.34 hydrogen, 1 hydrogen; and opposite 30.01 oxygen, 3 oxygen. Such a compound, then, will consist of three atoms oxygen, two atoms carbon, and one atom hydrogen [Prout 1815a: 270].

Although Wollaston slide-rules were made by all the London instrument makers, there is no evidence that Prout's modification was made professionally. It would, of course, have been a simple alteration to effect.

When Gay-Lussac (1815) introduced analysis by copper oxide in 1815 Prout quickly adapted the new technique to a charcoal furnace apparatus in which (unlike Berzelius's system) the combustion products were collected vertically (Prout 1817; Prout 1818a). His principal reason for abandoning the use of potassium chlorate as an oxidising agent (which Berzelius continued to use for several more years) was that any nitrogen present in the compound was released as the free element rather than as a series of confusing oxides of nitrogen. The new apparatus, which he described to the Medical and Chirurgical Society in 1817, consisted of a sturdy glass tube of $\frac{1}{4}$ inch bore and one foot in length which was filled with about four grains of the organic substance with an equal amount of copper oxide. Thin copper sheeting was wrapped around the tube to prevent accidents due to the cracking of the tube and excessive heating of the oxide, and the tube itself was arranged horizontally in an iron charcoal furnace with a Berzelius-type movable partition. Another tube, filled with calcium chloride to absorb water, led the combustion products to a mercury gasometer. All the substances for analysis were dried in Prout's vacuum apparatus beforehand, and the weighings were made with a delicate balance and weights which had been specially calibrated for him from platinum standards cast by the leading London instrument maker, Edward Troughton.

Apart from the gasometer (which had been personally calibrated by Prout) the apparatus's debt to Berzelius was clear. However, the apparatus described by Prout in 1820 was radically different. The major changes were that the combustion was conducted by a milder spirit lamp instead of a charcoal furnace and that the layout was vertical instead of horizontal (Prout 1820b). The circular spirit lamp, which enveloped the combustion tube, could be moved up and down by counterweights and the combustion products collected directly into a barometric gasometer. Other fittings designed for the apparatus in place of the gasometer enabled water vapour to be condensed and collected directly or for specific gravity determinations to be made. A Holborn instrument maker named Tuther evidently manufactured Prout's apparatus on a small scale and it was mentioned favourably by William Henry in his textbook *Elements of Experimental Chemistry* (1823: ii, 167–9).

French and German translations of Prout's paper describing his apparatus, as well as Berzelius's praise of it for 'ingenuity' in his *Traité de Chimie* (1829–33: ii, 33) made the spirit apparatus known throughout Europe and America. Although Prout himself claimed that his technique was an improvement on all previous analysts' apparatus, including his own, and that substances analysed by the method seemed to contain more carbon, he did not remain satisfied for long. By 1827 he had made a partial return to Lavoisier's technique of direct oxidation with a costly and highly elaborate piece of apparatus. His main reason for abandoning copper oxide (which other chemists did not follow) was his dislike of the fact that it readily absorbed both water vapour and air, thus making the remixing of the organic substance with the copper oxide half-way through the combustion—a procedure which Prout always insisted upon—a guaranteed method for introducing inaccuracy.

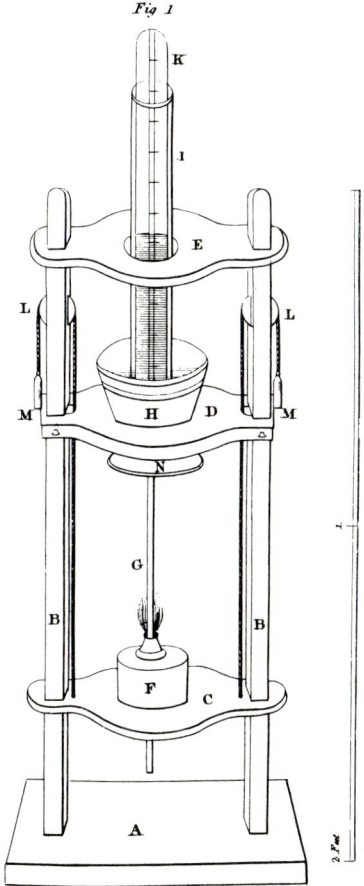

Prout's spirit lamp combustion apparatus for the elementary analysis of organic compounds (from *Ann. Phil.* **15** (1820) Plate CII).

> To conquer these [difficulties], every means that could be thought of, as likely to succeed, were tried, but without effect, and I was obliged to relinquish the matter in despair, and endeavour to contrive some other mode of analysis that should be free from these difficulties altogether [Prout 1827: 360-2].

Prout's complicated new instrument, which was probably constructed for him by Faraday's instrument maker, John Newman, used a spirit lamp to heat the horizontally arranged combustion tube. The latter was attached by stop-cocks to a gasometer burette. A measured volume of oxygen being admitted into one of these siphons, it was then transferred through the combustion tube to the empty gasometer, when the amount used in combustion was easily determined. In practice, Prout evidently experienced some difficulty in ensuring a complete oxidation of the organic substance and was forced, after all, to add a booster of copper oxide to the oxygen in the combustion tube.

If the addition of copper oxide destroyed the simplicity of the volumetric estimation, other disadvantages of the apparatus are readily apparent. Although no financial details survive, it must have been expensive to build. Large quantities of expensive mercury were required for its operation and it took even longer to make a determination than with the traditional time-consuming copper oxide method. Its accuracy, despite Prout's claims, was open to question (for example, any reduced copper might be reoxidised in the stream of warm oxygen) and, crucially, the apparatus as described could not be used for the determination of nitrogen. All of Prout's published analyses with this instrument were compounds of carbon, hydrogen and oxygen.

It was for this apparatus and the attendant analyses of sugars that Prout was awarded the Royal Society's highest honour—the Copley Medal—in 1827. In the words of Davies Gilbert (1828: 61), then President of the Society, the new analytical technique promised

> not merely to disentangle any one particular combination but to afford an insight into all the products created by living chemistry. They [i.e. the Council of the Royal Society] have hastened, therefore, to stamp with their highest mark of approbation, as well as the mode of analysis itself as the specimens of what, in the hands of Dr. Prout, it has already performed; and not doubting, but that by the exertion of such talents, such ingenuity, and such labour, their satisfaction will from year to year be continually increased.

Ironically, Prout was to publish no more analyses after 1827 and within three years, his search for the analysts' philosophical stone—the perfect technique of organic analysis—was achieved in Germany by Justus Liebig (1831). The latter returned to the charcoal fire of earlier analysts and by an ingenious triumph of glass-blowing weighed the products of combustion directly in bottles of lime water and anhydrous calcium chloride. Prout (1840: xcii) himself was unimpressed, though according to W C Henry (1837) he was 'anxious' to talk to Liebig personally about analysis.

> Liebig's analytical apparatus was in effect tried by me nearly twenty years ago, and for *rude approximations* it answers very well; but it is not, in my opinion at all adapted for obtaining *very accurate* results.

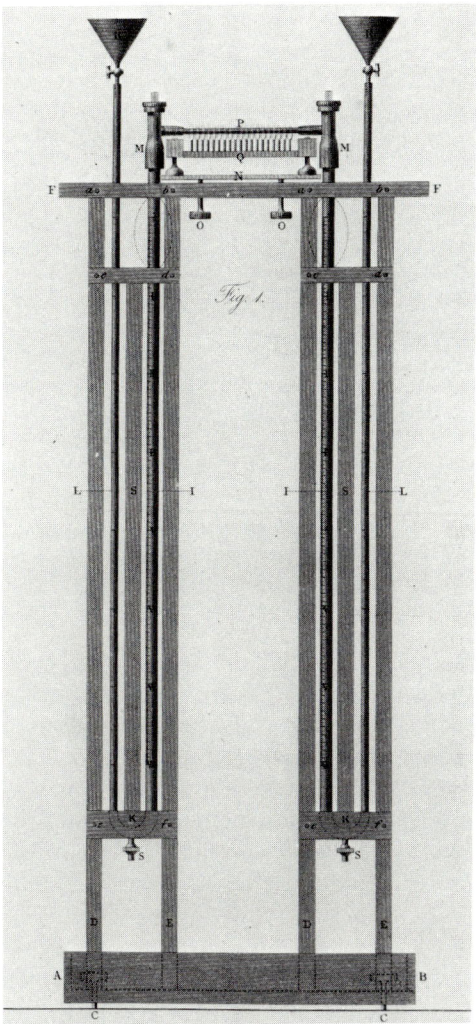

Prout's oxygen apparatus for organic analysis (from *Phil. Trans. R. Soc.* (1827) Plate XIV).

In one sense Prout was right, since for very accurate analyses a compromise between the techniques of Prout and Liebig has continued to be made, with a current of air or oxygen being blown over the heated copper oxide; indeed, Prout's particular apparatus continued in use in England for several years after the introduction of Liebig's method (Daniell 1843: 607; Brande 1848: ii, 1146). However, in a more fundamental sense, it is clear that Prout—who was so taken with the notion of the need for a simple theory of matter—failed to appreciate the need for simplicity as well as accuracy in analysis. This was where Liebig's method scored, and incidentally, by allowing any numbskull to crack the quantitative code of Nature's living chemistry, Liebig opened the floodgates of research into organic chemistry and biochemistry.

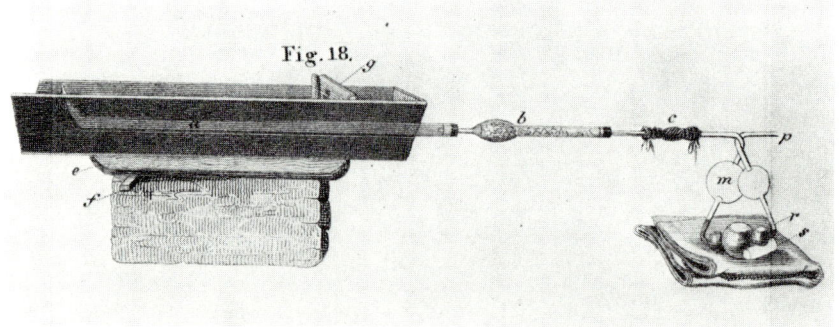

The apparatus for elementary analysis of organic compounds developed by Liebig between 1830 and 1831 (from *Anleitung zur Analyse organischer Körper* (1837) (Braunschweig: Vieweg)).

Too much attention has been paid in the past by historians of chemistry to the confusions and difficulties over nomenclature and atomic weights, vitalism and classification models in the development of nineteenth-century organic chemistry, and not enough consideration has been given to the tremendous technical difficulties with which the early organic analysts wrestled. But the purely intellectual problems of organic chemistry would never have been promoted or solved without the analysts' dogged search for adequate foundations. Each practical chemist searched for the analysts' philosophical stone. Prout spent twelve years of his life searching for an apparatus and a technique which could provide accurate analyses of organic materials, but he was never satisfied that he had solved the problem, nor did he ever, of course, tackle the problem of nitrogen estimation. Although Andrew Ure (1822) could write confidently that his own analytical techniques allowed him to complete six combustions a day, we can be sure that Prout worked at a much slower pace when it is remembered that he combined his chemical hobby with the duties of a medical practitioner. His friend, the Oxford chemist and agriculturist Charles Daubeny (1852: 99), aptly described the situation when he wrote:

> The greater part of Dr. Prout's analyses were made with an apparatus of his own which, however ingenious it might be, was far more difficult to use, and required for its success many more precautions than that at present [1852] in the hands of chemists, and hence the precision to which he attained is the greater subject for commendation. Add to which, that these delicate investigations were carried on by him, unassisted, amid constant interruptions, at intervals snatched from the daily demands made upon his time by professional engagements.

Urine Chemistry

Prout was already known in London medical circles as an 'animal chemist' when he first came to general scientific attention as an organic analyst with the publication of five short papers on the proximate analyses of various substances in 1815. Although these papers, for example on the ink of the cuttlefish (a polymer of indole which was far too complex for analysis at this time), were ephemeral and inconclusive, they all reveal his acquaintance with, and concern for, the improvement of chemical tests and analytical reagents, or show his general skill and ability as an analyst which was to be the subject of contemporaries' admiration. For example, in an Anglo-French dispute over the analyses of amniotic fluid, E W Brayley (1832), one of the editors of *The Philosophical Magazine*, accepted Prout's result because 'the confirmatory evidence' he had furnished was 'so valuable upon a point of this nature, on account of his minute acquaintance with animal fluids, and his practical skill in their examination'. Similarly, when Edward Turner, the first Professor of Chemistry at London University, began his programme of analyses to check Thomas Thomson's values for atomic weights, he used silver chloride as a standard substance. However, Prout, 'whose remarks are always pertinent, being founded on careful observation', pointed out to Turner (1833: 534; 1834b: 397) that 'pure' silver chloride usually gave out hydrogen chloride gas when it was dried, and that if this were not completely removed it could interfere with Turner's estimation of the atomic weight of chlorine. It was a concern for little details like this that was to play a major role in the testing of Prout's hypothesis throughout the nineteenth century.

Of far greater significance was Prout's series of papers relating to the chemistry of urine and digestion which originated in the analysis of a snake's excreta in 1815 and ended with the analysis of fossil faeces in 1829. A sample of the urinary faeces of a young boa constrictor on exhibition near the Exeter Change in the Strand was dissolved in hydrochloric acid and the insoluble portion shown to be uric acid by what is today recognised as 'the murexide test'. The addition of ammonium oxalate to the insoluble portion gave the additional information that lime (calcium) was also present (Prout 1815b). Quantitatively, to his obvious surprise, the excrement proved to be almost pure uric acid (90.16%). This was the first demonstration that uric acid, and not urea, was excreted by reptiles, though Prout recalled that W H Wollaston (1810) had found uric acid in bird droppings and that the guano of South America contained large amounts of the same substance. A few years later John Davy (1818), who had been practising medicine with the army in Ceylon, confirmed that uric acid was a constituent of the excreta of snakes and lizards in their natural habitat while Prout (1820a) himself confirmed it with a chameleon, crocodile and other reptiles. Although the discovery was to have important implications for theories of metabolism, its more immediate significance lay in palaeontology.

Although Prout also examined the boa constrictor's faeces, he did not publish the results. Samples were, however, given by him to Alexander Marcet (1817: 114) who concluded that they were composed largely from calcium phosphate, 'the undigested residue of the food of the animal'. Consequently, when the Oxford geologist William Buckland (1829: 223) wished to confirm his hunch that the coiled fossil stones called coprolites which he had found at Lyme Regis were a fossil faeces, Prout was asked for his opinion. The latter confirmed that coprolites were basically calcium phosphate mixed with smaller amounts of lime, carbon, iron and sulphur, and agreed that they resembled physically and chemically the faeces of living animals. Buckland had found abundant supplies of coprolites wherever there were fossil remains of icthyosauri, the fish-like marine reptile first identified by Cuvier, and both Buckland and Prout noticed bones of smaller fish within the faeces which were stained by a 'bright jet-black colour'. Here Prout's previous experiments with the ink of the cuttlefish came into its own for these colouring matters proved to be identical, thus providing crucial evidence that 'the Icthyosauri fed largely upon the Sepiae [cuttlefish] of those ancient seas'.

In the 1840s, with the rise of high farming in the British Isles, there was renewed interest in the calcium phosphate content of coprolites. In 1842, the wealthy landowner Sir John Lawes patented a process for the preparation of super-phosphate fertilisers and large areas of

Cambridgeshire were made over to the mining of coprolites (Partington 1964: iv, 313; Grove 1976). To see how Prout's demonstration that reptiles and birds were sources of uric acid led to another industrial development, we must return to his medical work on urine.

Prout had exhibited pure urea at his lectures on animal chemistry in 1814, and later, in two lectures to the London Medical and Chirurgical Society in 1817 and 1818, he presented detailed analyses of the primary constituents of both healthy and diseased urine. Prout's method of producing pure urea, which involved evaporating urine, separating the other constituents with nitric acid and purifying with animal charcoal, became the standard technique described in nineteenth-century textbooks. His figure for its 'proportional number' (or molecular weight) of 37.5 ($C = 7.5$, $N = 17.5$, $O = 10$, $H = 1.25$) was similarly uniformly adopted by other chemists, including Wöhler (1828) in his paper on the synthesis of urea (Turner 1834a: 871).

This analysis, together with those for urea nitrate, uric acid, sugar, diabetic sugar and lactose, not only convinced Prout of the applicability of Dalton's atomic theory to the products of living systems, but also seemed to confirm that they were all composed from 'the union of more simple compounds, as urea of carburetted hydrogen and nitrous acid, lithic [uric] acid of cyanogen and water, etc; circumstances which render almost certain that their artificial formation falls within the limits of common chemistry' (Prout 1817: 538).

The possibility of synthesis followed from the early radical theory of Gay-Lussac, according to which organic compounds were built up from simple radicals (groups of different elements which behaved like a single element), and analogous therefore to inorganic compounds. Although this was not the reasoning which lay behind Wöhler's experiments (Brooke 1968) it was the reasoning which led Prout to make several unsuccessful attempts to synthesise urea from ethane and nitric acid before 1820 (Prout 1848: 530).

Although Berzelius had initially, in 1814, been confident that organic compounds did obey the law of definite proportions, by 1818 he was suffering doubts. Hence, when Alexander Marcet told Berzelius of Prout's results in a letter, Berzelius (1912–36: i, 155) was sceptical:

> I am much obliged to you for sending me Dr. Prout's memoir, or at least the results of his experiments which he has printed. I will not hide from you, however, that I distrust most research of this kind on organic products in which one so easily finds simple and definite proportions; one is so easily deceived by the closeness of the results to such simple proportions when the true composition is, however, well removed.

As a result of these remarks, in February 1818, Marcet forwarded a copy of Prout's paper ('which seemed to me to have merit'). However, Berzelius (1912–36: i, 173) remained unimpressed by Marcet's claim that Prout's experiments had been made 'with much care and in good faith', replying that 'the results on the three animal substances obtained by Dr. Prout definitely did not merit the confidence that you have in them. Prout has not perceived the difficulties accompanying this problem and he is too confident in applying the law of definite proportions'. Once more Marcet sprang to Prout's defence: 'truly, you are doing him an injustice. He is a good worker and a true and modest man—though doubtless he can err like any other man. However, I believe him to be quite correct in what he has done' (Berzelius 1912–36: i, 174).

Although Berzelius made no reply to this letter, it seems likely that he was thinking of Prout's conclusions, as well as the work of Gay-Lussac, when in his *Essai sur la Théorie des Proportions* published in Swedish and French in the year following the controversy, he wrote:

> Some chemists have interpreted the composition of organic substances differently from the way I have just explained from my own experiments. They have been represented by the binary combination of inorganic elements. For example, the composition of sugar is represented as a compound of a volume of water vapour combined with a volume of carbon in the state of a gas [Berzelius 1819: 46].

But, Berzelius went on, how could the enormous number of organic compounds be explained in terms of such simple compositions? Even if Gay-Lussac's binary hypothesis was justified in the case of alcohol and ether, it could never work for sugar, which existed as three species 'whose differences, insofar as specific properties relate to composition, are inexplicable on this way of envisaging organic compounds'. If chemists adopted such an interpretation, they might fall into error by adapting their analyses to suit these preconceived views. Supposing there were different 'species' of organic compounds to be found in different animals? Thus, fibrin, albumin and urea might vary in some small way from one animal to another and yet be a single species within their genus. Indeed, he observed, the fibrin of a cow's blood did seem to be slightly different from human fibrin. So Berzelius's point was that chemists should not be too hasty in the application of rules of chemical proportions to organic compounds, for it might turn out to be as harmful to the progress of organic chemistry as it had been beneficial to the progress of inorganic chemistry. Until it had been shown that urea samples showed no significant variation, Prout's work was not, he concluded, to be taken very seriously (Berzelius 1912–36: i, 156).

Prout (1817: 541) also drew another conclusion in his 1817 paper:

The remarkable relation found to subsist between urea and sugar, seems to explain in a very satisfactory manner the phenomena of diabetes, which may in fact be considered to consist in a depraved [i.e. pathological] secretion of urea. Thus the weight of the atom of sugar (18.75) is just half that of urea (37.5): the absolute quantity of hydrogen in a given weight of both is equal, while the absolute quantities of carbon and oxygen in a given weight are precisely twice those of urea.

Such analyses seemed to afford Prout with 'glimpses of laws that will hereafter be found to influence the whole system of Nature's operations'. This Pythagoreanism, or fixation with chemical arithmetic, which first found outlet in his anonymous papers of 1815–16 (Chapter 5) received amplification in a second paper on the chemical constituents of urine (Prout 1818a). Its purpose was to show the relationship between the different constituents of urine and the albumin of the blood. A sample of the latter was taken from one of John Elliotson's diabetic patients. Although it could not be assigned a rational composition, the percentage composition was, in agreement with earlier analyses by Gay-Lussac and Thenard, the same as for the albumin of healthy blood. However, the urine of this diabetic patient had a nitrogen content some 28 per cent *less* than normal urine. Here the significance of Prout's generalisation becomes plain: 'Diabetes is a depraved secretion of urea' in the sense that urea was perhaps secreted as two molecules of sugar instead of as one molecule of urea. Such an explanation, of course, made the transmutation of nitrogen necessary—a possibility which, as we shall see, Prout did not find implausible.

Prout's analyses of 1817.

Elements	Urea		Sugar		Uric acid	
	atom	per cent	atom	per cent	atom	per cent
Hydrogen (1.25)	2.5	6.66	1.25	6.66	1.25	2.85
Carbon (7.5)	7.5	19.99	7.5	39.99	15.0	34.28
Oxygen (10)	10	26.66	10	53.33	10	22.85
Nitrogen (17.5)	17.5	46.66	—	—	17.5	40.00
	37.5	100.00	18.75	100.00	63.75	100.00

Since publishing his first paper on urine in 1817 Prout had found several imperfections in his analytical method and he therefore went some way towards justifying Berzelius's criticism. 'A more extensive experience in the analysis of organic substances', he admitted to his medical audience, 'has made me acquainted with various circumstances which I never suspected, and which are of the utmost importance in this most difficult branch of practical chemistry' (Prout 1818a: 478). Nevertheless, despite a serious admission that he had not overcome these difficulties sufficiently to leave no doubts 'in any one given experiment', he tabulated his results and boldly drew readers' attention to 'the extraordinary relations that exist among the ... numbers'.

Prout's analyses of 1818.

100 parts	H	C	O	N	Weight of one atom	Sugar = 1
Sugar	6.66	40.00	53.33	—	18.75	1
Urea	6.66	20.00	26.66	46.66	37.50	2
Uric acid	2.22	40.00	26.66	31.11	56.25	3
Cystine	5.00	30.00	53.33	11.66	75.00	4
Oxalic acid	4.44	20.00	75.55	—	112.5	6
Albumin (from blood serum and egg white)	7.77	50.00	26.66	15.55	112.5	6

From this table it would appear that the molecular weights of sugar, urea, uric acid, cystine (where the presence of sulphur was undetected) and oxalic acid formed an arithmetical series, or were related by multiples of the weight of sugar. Such a relationship appeared even more significant if albumin from blood serum or an egg white was included in the sequence. Here was the beginning of the metamorphosis theme of Prout's later molecular and physiological studies, wherein molecules underwent 'reduction' and 'completion'— in our language, depolymerisation and polymerisation—to the orchestration of 'organic agents'. However, this theme is best postponed for consideration until 'Prout's hypothesis' has been examined in Chapter 5.

Prout's work on urine, for which he designed a hydrometer for use in clinical practice (Prout 1825b), led him to the discovery of a substance he and W H Wollaston named purpuric acid. This research formed the subject of Prout's first communication to the Royal Society, to whom it was read by Wollaston on 11 June 1818. No doubt his intention was to gain election to the Society for this happened soon after the paper's publication.

The purple colour produced in the reaction between uric and dilute nitric acids had first been described by the Swedish pharmacist, Scheele, in 1776. Prout had used the coloration as a test for the presence of uric acid in a reptile's faeces in 1815; he now ascribed the colour to the ammonium salt of purpuric acid, so named because of 'its remarkable property of forming compounds with most bases of a red or purple colour'. The free acid proved to be a fine yellowish powder, though as Prout came to realise with the publication by Liebig and Wöhler of their comprehensive work on uric acid derivatives in 1838, it was not really a free acid he had prepared but an ammonium compound of a purine base which Liebig named *murexide*. This, Liebig and Wöhler (1838) found, rapidly hydrolysed to form a mixture of what they called *alloxan* and *murexane* (modern uramil). The latter they identified with Prout's purpuric acid. In the later editions of his treatise on urinary diseases. Prout accepted Liebig and Wöhler's account of these purine derivatives, though he protested about the 'barbarism' of their 'extraordinary' names.

The bulk of Prout's paper on purpuric acid was devoted to a description of a variety of inorganic purpurates. From the brilliance of their colours he was led to speculate (correctly) that such salts might be the basis of many animal and vegetable colours. If this were the case, he pointed out, then they and the reptilian and avian uric acid from which they could be prepared, could be of considerable artistic and commercial value. Indeed, when guano began to be imported from South America in the 1840s, it was used not only as a fertiliser but as a source of murexide dyes by dyestuffs chemists in Manchester and Mulhouse in France. Ironically, Perkins' purple dye, derived synthetically from aniline, which heralded the demise of the murexide dye industry, was initially marketed in 1856 as a substitute (Mellor and Cardwell 1963; Homberg 1983).

Another curious new acid was announced by Prout in 1822 when Alexander Marcet read a paper to the Medical and Chirurgical Society on 'a singular variety of urine which turned black soon after being discharged'. Marcet had obtained the sample from William Babington, the physician and friend of Sir Humphry Davy, in December 1814, the patient being a child of 17 months who must have been suffering from the metabolic disorder known as alcaptonuria. Marcet had lost contact with the child, but passed its urine samples to Prout for analysis. The latter, who could find no urea or uric acid in the urine, reported that the black colour was due to an unknown principle combined with ammonia which he proposed to call, appropriately, melanic acid. This was homogentisic acid (2:5-dihydroxyphenylacetic acid). Although chemically quite distinct, it seems strange that Prout saw no analogy between this acid and the ink of the cuttlefish, melanin, which he had previously examined.

Much more common than alcaptonuria in Prout's day were the urinary diseases in which the metabolic disorder was signalled by the presence of painful stones or calculi in the kidneys and bladder. His growing reputation as a physician who specialised in urinary complaints inevitably led him, like Marcet, to take an interest in the chemical composition of such stones. In 1819 he detected ammonium urate in large quantities in a few calculi, thus repudiating the conclusions of recent authorities such as Brande (1808), Henry (1819) and Marcet (1817) that earlier French claims for its existence were erroneous (Prout 1819b).

All of Prout's early work on the chemistry of healthy and pathological urine was elaborated into a little book published in 1821 which 'established his reputation as a chemist and practical physician' (Munk 1878: iii, 109). This *Inquiry into the Nature and Treatment of Gravel, Calculus and other Diseases of the Urinary Organs* was one of the first nineteenth-century medical textbooks to deal with disease from a chemical viewpoint. An instant success, its merits are well measured by the fact that it was translated into French and German—a mark of distinction at a time of continental supremacy in medicine, when translations usually went the other way.

The *Inquiry* opened with a chemical comparison between the constituents of human blood and urine which served to establish a norm from which diseased urine deviated. It also allowed Prout, in a later edition, to draw the significant conclusion for an understanding of metabolism that urea was first formed in the blood and subsequently secreted through the kidneys by an organising agent (Prout 1848: 529, 531). Although he noticed the presence of albumin in pathological urine, until Richard Bright correlated its presence with dropsy and renal disease in 1827, it was not known that albuminuria was a specific signal of kidney failure. Not surprisingly, therefore, Prout did not offer his readers any explanation for albuminuria, or suggest any treatment. However, from the second edition (Prout 1825a) onwards Prout always distinguished between *chylous* and *serous* albuminous urine; the former kind of diseased urine contained a form of albumin which resembled the incipient albumin of chyle (hence *chyluria*), and the latter contained an albumin identical with that found in blood serum (*albuminuria*). 'Distinctly defined instances of both these varieties' were rare and usually urine was found to contain both varieties of albumin. In the final edition of his textbook (Prout 1848) he renamed the former *chylo-serous* urine in order to differentiate it from the *serous* urine made familiar by Bright's work. As we now know, the symptoms of both chyluria and albuminuria are well-defined. While the latter frequently occurs independently of the former, chyluria (which is usually caused by a lesion in the lymphatic system which allows fats from the digestive system to pass into the urine) is not dependent upon a kidney lesion. On the other hand, Prout's term chylo-serous urine is quite appropriate since in chyluria albumin will pass into the urine along with fats. Prout correctly noticed the intermittent character of chyluria and that it was more common in the tropics than in Europe; the parasitic cause of chyluria, and its connection with the lymphatic system was only established at the end of the nineteenth century.

In his discussions of the diseases of urine, Prout adopted a classification based upon the solubilities of the different chemical substances he found in pathological urine. Soluble principles such as urea, sugar and albumin could be morbidly deranged in both quality and quantity; insoluble principles appeared as gravels and calculi. In order to speed diagnosis, Prout published a routine for testing the urine which could easily be performed by the general practitioner in the patient's home. For example, he gave the following test for detecting excess urea—a symptom of chronic nephritis.

> The modes which I commonly use to detect an excess, is to put a little of the urine into a watch glass, and add to it carefully nearly an equal quantity of pure nitric acid, in such a manner that the acid shall subside to the lower part of the glass, from its greater specific gravity, and allow the urine to float above it. If spontaneous crystallization takes place [urea nitrate], an excess of urea is indicated; and the difference of excess can be inferred, near enough for practical purposes, by the greater or less time which elapses before crystallization takes place, which time may vary from a few minutes to two or three hours [Prout 1821: 10].

Detailed 'practical rules for determining the nature of the Affection and its appropriate Remedies, from the properties of the urine and other Symptoms' were published in the second edition of his treatise in 1825. In order to detect variation, he pointed out, there had to be a standard examination procedure.

> I prefer [to use] a transparent cylindrical vessel, such as a common phial, of not less than one inch, not more than two inches in diameter, and from six to eight inches long. In such a vessel all the sensible properties, both of the urine and its deposites, can be distinctly ascertained [Prout 1825a: 286].

Prout listed the necessary kit of apparatus as: red and blue litmus paper; turmeric paper; a watch glass or thin platinum disc (for evaporation); two small discs of plate glass for distinguishing pus from mucus by an optical test devised by Thomas Young; a specific gravity bottle or portable hydrometer; blowpipe and forceps for tests on gravel and calculi; test tubes; stoppered phials containing solutions of reagents such as ammonia, potash and nitric acid. All these materials 'can be readily packed into a small portable case, or pocket book, and will be sufficient, by the aid of a common taper, or candle, to perform all the experiments on the urine, and urinary productions, that are commonly necessary'. It is interesting to note that Prout did not need to mention this chemical kit in subsequent editions, presumably because the techniques he was describing had become part of standard clinical practice by 1840.

Urine examined first thing in the morning, Prout believed, would best determine the kidneys' health; urine examined after a meal would best reveal digestive disorders. The physician was required to note fluctuations in the amount of urine excreted, its colour or transparency, its specific gravity (the average range was 1.01 to 1.015) and its acidity and alkalinity. For acidity Prout used ordinary litmus and the following precise test for uric acid.

> A very delicate test for the presence of a *free* acid in the urine, is the precipitation of lithic [uric] acid from it in the solid state, and the quantity of free acid present may be commonly judged of pretty nearly from the *time* required to produce this effect, and the quantity of lithic acid precipitated [Prout 1825a: 292].

Curiously Prout did not recommend a murexide test here, although he noted that the presence of urinary sediments, such as the urates and alkaline earth phosphates could be detected by the colour of the urine. The first edition of his treatise even carried a hand-painted endpiece which was designed to show the typical yellow, pink and red sedimentation colours of uric acid diathesis. (In practice the illustration was not very helpful since the colours were variable between copies.) He also appears to have examined urinary deposits under the microscope from about 1820 onwards; but he only recommended the general adoption of microscopy in 1843 after the appearance of the achromatic microscope (Prout 1843: plate).

Urine analysis is now a familiar, simple but important, part of medical diagnosis; to a very large extent it was Prout who brought uroscopy back into general medicine, placed it upon a strictly observational basis and stripped it of the medieval mysticism with which it had been once associated. 'I cannot too strongly impress upon my readers the necessity of frequently examining the state of the urine', he declared.

> Those who wish to know anything respecting the deranged operation of the urinary organs must submit to this drudgery, or be content to remain ignorant . . . Patients should also be directed to make general observations upon this subject themselves. In particular, they should be directed to keep two or three large wine or ale glasses in their bedroom, and observe the state of the urine at different times of the day, especially in the morning and evening, and likewise to note the changes which it undergoes by standing for some time [Prout 1821: 206].

In the medical section of the previously mentioned paper on the constituents of urine which he read to the Medical and Chirurgical Society in 1817, Prout had mentioned that he had practised uroscopy for several years before meeting Dr Charles Scudamore (1779–1849), 'who I found entertained similar views and had prosecuted the subject much further than I had done'. Scudamore, the author of a *Treatise on Gout and Rheumatism* (1816), gave Prout morbid urine samples from his own patients, and the discussions with an older expert must have been helpful. Unlike Scudamore and Marcet, Prout refused to suggest that there was a simple chemical remedy for the stone; in his view the only guaranteed solution was the knife for 'when a calculus is once formed, a further enlargement is probably a common chemical process, and will proceed whether the urine be healthy or not, for all the urine naturally contains the ingredients most commonly met with in calculi' (Prout 1817: 549). Of chemical remedies he was completely sceptical, both because of their potentially dangerous side-effects, and because 'the object of the chemical practitioner is at best but of a secondary description, namely to prevent the effects of diseases rather than to remove it [*sic*]' (Prout 1821: 206; Coley 1971).

Prout's determination 'to confine his attention chiefly to practical points' remained paramount in all five editions of his book. He further confined himself to what he called 'illustrations' and left the establishment of his conclusions to medical colleagues. Since his object was 'the truth', he wrote somewhat dramatically, 'whoever will direct him to this object, where he has failed to reach it, will be esteemed a friend'. A second edition of the *Inquiry* which appeared in 1825 included a handsome pull-out sheet of hand-coloured engravings by Francis Lunn of the different types of urinary calculi and material drawn from the work of the German, S T von Soemmerring, on the diseases of the kidneys and bladder.

In 1840 Prout published a third edition of his treatise under the new title *On the Nature and Treatment of Stomach and Urinary Diseases*. Although it was supposed to be a revised edition of the *Inquiry*, it would more correctly be described as a different book. Munk (1878), the compiler of collective biographies of Fellows of the Royal College of Physicians, believed that its publication marked an era in the history of animal chemistry, while Prout's Oxford friend, Charles Daubeny (1852), noted how Prout had 'abstained as much as possible from ... speculations, and [had] evinced an exemplary caution in confirming his practical deductions strictly within their legitimate limits, at the same time that he [had] displayed a profound sagacity in the discrimination and treatment of the diseases which fell within his province'. To the modern reader the book's interest is diminished by Prout's practical bias and his lack of theoretical discussion. Indeed, with the two further editions which followed in 1843 and 1848, even contemporary reviewers criticised Prout for not examining and explaining some of the theoretical issues involved in physiology (Anon 1841). Clearly it was Prout's deliberate intention to avoid controversy and we should not overlook the fact that such a practical emphasis may have been welcomed by contemporary general practitioners. Prout himself ignored his critics and made it clear that the only criticism he would listen to would have to come from 'the experienced chemical pathologist' (Prout 1840: Preface).

Historically, the most important and interesting section of the book is the introductory first part which contains a succinct and original non-Liebigian account of the physiology of the digestive and urinary systems—much of it, as we shall see in the next chapter, a specialised version of material previously presented by Prout in the third section of his *Bridgewater Treatise*. Yet, in the two remaining editions, Prout unwisely placed this biochemical introduction at the end of the volume as a third book, as if to emphasise still further the work's practical character (Anon 1848b). It was, however, impossible to understand the first two books, which employed his own special vocabulary and classification, without referring to this section. It was as if Liebig were to have published his revolutionary *Animal Chemistry* in 1842 with all the speculative chemical equations at the end of his book.

In this chapter we have been concerned to examine Prout's analytical results in the field of urine chemistry between 1814 and 1840 which followed from his efforts to invent an accurate method of organic analysis. In this field he made a number of significant discoveries for both organic chemistry and biochemistry: the discovery of ammonium urate in the urine of the boa constrictor eventually led to the generalisation that birds and reptiles differed from mammals in that the end product of protein metabolism was uric acid and not urea; the discovery of purpuric acid might be said to have laid the foundations for the investigation of purine and its derivatives, and on a personal level it led to Prout's election to the Royal Society. Such work was never divorced from his chosen career as a London physician and, in flowing from it, led him to rehabilitate uroscopy as a diagnostic technique. Despite his obsession with chemistry, however, Prout was sober enough to realise that urinary diseases would never be cured by chemotherapy alone. Instead of administering chemical, and perhaps harmful, remedies, Prout approached the bedside like an Hippocratic physician, emphasising the pathological importance of urine in diagnosis by recommending simple chemical procedures which could be used by any practitioner for analysing the urine. His experience was presented concisely and systematically in book form in

1821, and this gave medical practitioners a complete guide to the whole subject and made the author the first authority in London on diseases of the urine.

As we shall see more clearly from the next chapter, while he worked on urine chemistry he was at the same time investigating the chemistry of digestion and assimilation. In 1840 he welded together these two fields of research into a massive textbook which served as a complete guide to normal and pathological metabolism until it was replaced by the more powerful insights and resources of Liebig's school.

However, Prout was not concerned throughout his life with the mere accumulation of chemical facts; always he sought for the laws which governed 'not only the operations of the animal economy, but the whole material world'. The Daltonian atomic theory, as modified by Berzelius, seemed to be a promising way of rationalising the complexities of organic chemistry, and already in the papers of 1817 and 1818 it can be seen that Prout not only felt that organic synthesis was possible, but also that the ultimate explanation of physiological chemistry might lie in a theory of molecular rearrangements or even transmutations of the elements. With this in mind we turn to consider the other principal subject of his researches, the chemistry of digestion.

3

Physiology

> With respect to the subject I am aware that it is purely metaphysical and for that I should offer an apology were I not convinced that my professors will pardon me when they consider how closely metaphysics and physiology are connected [Prout 1810a].

In September 1813 Walter Adam, who wanted to become a surgeon, sought Prout's advice on which courses to take at the University of Edinburgh. In reply Prout (1813b) strongly recommended physiology, as well as chemistry. 'Physiology is a noble study. It is of all others to me the most agreeable.' Apart from the dissertation of 1811, Prout published nothing until 1813. However, a review note on progress in physiology written by a 'friend' of Thomson for his *Annals* discussed the theory that teeth were analogous to hair and nails, and mentioned that:

> A similar opinion was advanced in 1811 by Dr. Prout, who at that time drew up the sketch of a paper, the object of which was to prove that the teeth are to be considered as appendiges to the integuments, and to be classed with horns, nails, &c. The opinion was principally founded upon extensive anatomical inquiries, showing the analogy between the formation of the teeth, horns, feathers, &c, and partly also upon physiological and pathological reasonings. The paper was never published, owing to reasons which need not be mentioned, but the opinion was stated to many of the author's friends at the time [letters to Adam confirm this], and he intends at some future opportunity to lay the subject before the public in extended form [Anon 1820b].

Needless to say, no paper on the subject was ever published by Prout, though a paper written in June 1813 confirms that his earliest interests were physiological. The problem which Prout (1813c) had examined was how to stain the blood vessels of anatomical specimens by the old art of anatomical injection (Cole 1921). After many trials he had found that a saturated solution of potassium ferrocyanide and dilute ferric sulphate would successfully plant prussian blue in the morbid tissues. The injections were made through a syringe after the solutions had been heated to 100 °F. In an investigation of the vascular nature of the ox's eye, Prout found that 'the vessels of all parts of this organ appear to communicate freely with one another; the part least connected with the rest is the retina, and this is supplied by its own proper artery'. He also successfully stained the lens capsule and decided that the hyaloid membrane 'in the adult state at least ... derives all its vessels from the great arterial communication situated a little behind the ciliary ligament, and not from the retina, as usually stated'.

This promising line of research was probably not pursued because, as Prout hints, he experienced social difficulties in using morbid materials and in dissecting in his private apartments. Instead, he switched to self-experiment in 'observations on the quantity of carbonic acid gas emitted from the lungs during respiration, at different times, and under different circumstances' (Prout 1813a). His intention here was to determine both whether the quantity of carbon dioxide exhaled in the breath was constant throughout the day, and constant for the individual. The analyses were made on himself with the aid of a breathing apparatus similar to the modern spirometer and adapted from a system used by Allen and Pepys (1808). A strict regimen 'which consisted in keeping myself as nearly as possible in the same state in every respect' was adopted. He stuck to this 'arduous' discipline for nearly three weeks during August 1813, 'making experiments every hour, and sometimes oftener, during the day, and occasionally during the night also'. Prout (1813a: 329) thought he perceived a pattern in his results, stating as a law:

> The quantity of oxygen consumed, and consequently of carbonic acid formed during respiration, is not uniformly the same during the 24 hours, but is always greater at one and the same part of the day than at any other, that is to say, its maximum occurs between 10 a.m. and 2.0 p.m., or generally between 11.0 a.m. and 1.0 p.m.; and its minimum commences about 8.30 p.m., and continues nearly uniform till about 3.30 a.m.

No such law will be found in modern textbooks of biochemistry, for Prout was partly misled by the severity of his regimen; he tended to underrate the influence of food, or rather, the importance of noting the times of his meals and fasts. The correlation of all the possible variables which may influence metabolism only began to be fully unravelled in the magnificent research programme of the nutritionist

Edward Smith (1859) half a century later (Chapman 1967). It was Smith who found that there was a maximum output of carbon dioxide some two hours after a meal, a corresponding minimum before meals, and a 'basal or normal' state when the body was at rest. Obviously Prout's absolute minimum for the early hours of the morning represented Smith's basal state (or basic metabolic rate, as it would now be termed); while his noon maximum was probably due to his one substantial meal of the day.

Since his frugality actually led him to rule out food as the source of variations, Prout (1814a) looked elsewhere for an explanation in the presence or absence of sunlight. This itself was not entirely unreasonable, for there is a seasonal variation in metabolic rate; but by insisting that the sun alone was responsible, his results became metaphysical rather than experimental.

Prout's speculative streak had already been strongly present in a fascinating essay on sensation which he had written as a student and it was to reappear in the 1820s in his writings on digestion.

Sensation

De Facultate Sentiendi is a quarto manuscript of 24 pages, dated Edinburgh 1810, in which Prout promised 'to deliver my own opinions only, respecting the curious and interesting subject of sensation, without paying much attention to what has been said by others on it'. A clumsily written, but important, undergraduate essay, it not only provided Prout with some material for his first published paper (Prout 1812), but it also throws much interesting light on the origins of what later became known as 'Prout's hypothesis'. From the manner in which he addressed 'my professors' in the essay it would appear to have been a memoir written to satisfy the demands of the medical course. It has nothing to do with the dissertation submitted for his medical degree in the following year, and the best modern analogy would be with the 'project' essay which is set in many universities, and of which several examples exist in Edinburgh University library for earlier and later periods than Prout's stay in Edinburgh.

Prout began this essay by suggesting that the nature of the vital principle which organises vegetable and animal principles (i.e. organic compounds) into living creatures remains an inscrutable mystery simply because there was nothing with which it could be compared in quality or degree. 'We are obliged to conclude that it is a principle *sui generis* or first principle' (Prout 1810a: ii), and to define it indirectly from its observed effects.

> The living principle is that principle which when combined with matter has *apparently* the power of imparting to it one or more of the three following properties viz *vegetation, sensation,* and *capacity for knowledge,* according to the different modes or degrees in which it combines with it.

These three properties, or categories, correspond to the three Aristotelian souls, vegetable, animal and intellectual, which exist together in man alone. Prout calls their respective combinations with matter 'vegetative', 'sensitive' and 'intellectual combinations'. This classification was to form the metaphysical foundation for Prout's later published views on nutrition. The function of the vegetative combination is never mentioned in the essay, and we can only suppose that it was introduced in order to square with traditional physiology; otherwise Prout could find no analogical use for it. The intellectual combination, which is only found in man, is especially important since it has the power of acquiring knowledge 'through the medium of a property of the sensitive combination termed sensation' (Prout 1810a: 1).

What is the nature and subject of this knowledge, and how is it acquired by the intellectual combination? Prout argues that the capacity of the intellectual combination for acquiring knowledge is compounded from three Lockean properties of this combination: perception, memory and reason. If sensation is preliminarily defined as the 'property of the sensitive combination which is called into action when matter is brought into contact with any part of the sensitive combination or body of a sentient being', then perception may be defined as 'that faculty of the intellectual combination which furnishes the basis ... or the materials through the medium of sensation with which memory and reason operate in the formation of knowledge'. Memory therefore becomes the faculty of the intellectual combination which recalls without perception what it has previously conceived; and reason another faculty through which the intellectual combination can select and apply perceptions which have been retained by the memory.

But what is *matter*? That is, what is our knowledge of matter? Prout's answer is to take a definition from the *Philosophical Arrangements* of the eighteenth-century Scots classicist, grammarian and pedant, James Harris.

> Matter is that elementary constituent in composite substances which appertains in *common* to them all without distinguishing them from one another [Harris 1775: 63; Prout 1810a: 2].

By this definition all matter is of *one* substance, the primary matter which early Greek philosophers had distinguished as τὸ ὑποκειμένον. However, this first matter obviously exists in many conditions or forms of secondary matter: the ὕλη πρώτη, or 'matter which has a *capacity* for becoming *many* things before it *actually* becomes any of them' (Harris 1775:72; Prout 1810a:3). Prout suggests there is an analogy between the first matter and the uncombined living principle and a further analogy between secondary matter and the first combination of the vital principle with matter (the intellectual combination) since if perception could take place any way other than through sensation, this is the only condition of matter which could be perceived. A third condition of matter is the 'physical' which is endowed with the primary qualities of extension and hardness. Physical matter is analogous to, and adapted to, the sensitive combination, and it induces sensation. Prout emphasises that none of these kinds of matter has a real existence in nature; that is, at the empirical or chemical level they are confined and cannot be isolated. We can only arrive at a knowledge of these categories by processes of abstraction and analogy.

However, there does exist in nature an aggregated matter which enables the intellectual combination, through its faculties of reason and memory, to deduce a knowledge of such primary conditions of matter. In fact, Prout's whole purpose in making these distinctions 'is merely to endeavour to render probable by their means the unity of matter' (Prout 1810a:3). The ὕλη πρωτη, or secondary matter (we shall call it *protyle* in the sequel), is then redefined by Prout as 'matter in its aggregate state', by which he means such empirical materials as wood, stone, water, air and, presumably, the chemists' elements. Secondary aggregated matter may exist in five phases or physical states which depend upon the state of aggregation: solid, liquid, aeriform (gas), etheriform and luciferous. This semi-empirical classification now becomes the basis for much extravagant analogical reasoning.

How does the secondary aggregated matter differ from the final form of primary physical matter? Prout argues somewhat incoherently that the sole difference lies in a sensible secondary quality called roughness which is added to the primary sensible qualities of extension and hardness. Perhaps the difficulty in following Prout in this step arises from the use of the terms primary and secondary qualities in the context of Aristotelian physics. However, in this he was merely following Harris (1775: 88) who, although an Aristotelian, had briefly mentioned the Lockean language of primary and secondary qualities. Prout seems to have conceived that extension and hardness were perceptual qualities of matter itself (i.e. primary qualities); while roughness was a secondary quality whose perception depended on the state of aggregation of matter. Only after this preliminary metaphysical analysis was Prout able to discuss the manner in which the intellectual combination determines the existence of matter by means of sensation.

Matter excites sensation in the sensitive combination through the primary, but relative, qualities of extension and hardness. When a finger is drawn along the edge of a table a series of perceptions is received through sensation, and the faculties of memory and reason enable comparisons to be drawn between these sensations and other previously experienced sensations so that it becomes 'known' that matter exists 'at every point between the angles of the table and hence that these two angles could not be the same angle' (Prout 1810a:5). Without the aid of memory and reason, perception alone could not lead us to a knowledge of the extension or of the hardness of the table. But without sensation there would be no knowledge at all.

When matter makes contact with any part of the sensitive combination of a sentient being, a sensation will be experienced by it, and perception in the intellectual combination will follow 'though in different degrees of intensity &c. according to the acquisitness [sic] &c. of the organisation of the part in which contact takes place'. The succession, contact followed by perception, is extremely rapid, and the process is easily, but erroneously believed to be one of cause and effect.

Prout's opinions on the mechanism of sensation are then advanced in the form of two propositions. Firstly, since sensation is an effect (or property) produced by the combination of the living principle with matter, and since the cause of the effect is simply matter which is really only of one kind, it follows by analogy 'that sensation is but of *one* kind also'. Furthermore, since there are different kinds of secondary aggregated matter, sensation forms 'the basis or substratum of all those varieties of it which apparently take place when different conditions &c of matter are brought in contact with different parts of the bodies of sentient beings'. In fact there are five conditions of matter. Therefore, by analogy, there will be five kinds or modifications of sensation: touch, taste, smell, hearing and vision. Secondly, matter in its aggregate solid form is made up of secondary aggregated particles which are themselves formed from the primary physical particles (extension and hardness) combined with the secondary sensible quality of roughness. But roughness is distinguished by the modification of sensation, touch, and hence (by analogy again) all other forms of sensation and their variations in individual intensity will also depend upon roughness. Now the degree of roughness is dependent upon the sizes of the aggregated particles of secondary matter. Therefore, Prout concludes, tastes, smells, sounds and colours all ultimately depend 'upon the different sizes &c. of the aggregated particles of the same matter' (Prout 1810a: 8).

Certain parts of the body, like the lips, glans penis and the finger tips are more sensitive than others, while other parts are especially contrived so that 'the contact of external aggregated matter ... is effected through the media of certain mechanical apparatus'. These devices, the tongue, ear, eye and nose are so special that the sensations produced are dramatically modified 'and appear altogether different from sensation in its usual form' as experienced through the skin or lips. For example, the eye is adapted to the subtle form of aggregated matter called light which, by contact with the bare optic nerve, 'produces that modification of sensation termed seeing'. It will be noticed that Prout does not differentiate between general sensory perceptors, like the mucous membranes, which are susceptible to chemical irritation, and the more sophisticated localised receptors which he tabulates as follows:

Condition of matter	Part of body affected	Senses	Sensation
aggregated solid	fingertips	touch	feeling
liquids	tongue	taste	savour
aeriform	nose	smell	odour
atheriform	ears	hearing	sound
luciform	eyes	vision	colour

Prout makes seven comments on this table.

(1) Analogy leads to the following conclusion. Since physical matter, if it could exist in isolation, could produce sensations and perceptions through its sensible qualities of extension and hardness, 'these must necessarily be composite ... that is to say they must consist of sensation and perception of *matter* combined with certain qualities corresponding to those which exist in conjunction with the matter itself & these qualities of sensation & perception I shall term *quantity* & *intensity* supposing the first to correspond with extension & the second with hardness, &c'.

(2) Roughness—if the secondary particles of aggregated matter are all equal in size, the sensation of roughness is uniform; if they are different in size, the roughness is irregular. Furthermore, 'this modification of sensation like all others is varied in two ways: by the original sensible qualities of physical matter, viz. in quantity & intensity ... the quantity of the modified sensation corresponds with the quantities of surface acting at the same time in its production, while its intensity corresponds with the degree of roughness (whether greater or less) of the superficies of the same matter acting at the same

time in producing it'. The modern analysis of touch is more complex for we now recognise three receptors, touch, pain and temperature.

(3) Gustation is only excited by matter in the liquid state in which the cohesion of the secondary particles is much less than in the solid state. Taste is appreciated by the tongue and 'varies in quantity & intensity ... corresponding to the quantity of matter acting at the same time, the latter to the different tastes'.

(4) Olfaction is produced by the action of the aeriform aggregation of matter on the nose. In this state the particles are too small to be appreciated by either the body or the tongue as touch or taste. Smell is also varied in quantity and intensity corresponding to the quantity of matter acting at the same time, and to different odours.

(5) Hearing is only excited in the special organs, the ears, by a very subtle aggregated matter which may or may not be the 'aether' of the physicists. Hearing varies in quantity and intensity, but these are more usually termed loudness and acuteness and gravity.

(6) Vision is only excited in the eyes by the most subtle form of all aggregated matter, light. 'Vision varies in quantity & intensity the latter answering to the variety of colors the former to the brightness'.

(7) In a note Prout (1810a:13v) suggests the probability that the nerves which serve the various sensitive instruments, the tongue, nose, ear, etc, are also specially adapted to respond only to a specific state of matter, 'but if we may judge from the analogy of the unity &c of the acting matter this adaptation must consist only in different degrees of fineness &c of organisation in them'.

Clearly, Prout's purpose in *De Facultate Sentiendi* was to reduce all sensory mechanisms to a unity. In an explicit recapitulation he stated:

> In order to set forth the [speculation, *deleted*] probability of the *unity of matter* as distinctly as possible we have begun by endeavouring to trace its identity through all its primary conditions from the ὕλη πρώτη of the ancients down to its physical state, or that condition of it in which it becomes the object of sensation from its being vested with sensible qualities. *Respecting this nature of the unity of matter*, I may here observe that it was adopted by most of the ancient philosophers, and by many of the moderns among the latter of whom was I believe the immortal Newton himself. The speculations still are by many accounts visionary, but when we reflect upon the astonishing discoveries that have been made in chemistry and the progress it is still making, *who will say that at some future day they will not be realized at least with* matter in its secondary or aggregated condition [Prout 1810a:14].

Although the programme of reducing the number of chemical elements was to be placed upon an experimental basis by Prout a few years after he wrote these words, his programme to reduce the number of sensations has never been realised. In searching for analogies Prout overlooked significant differences, and he often fell into the simplest pitfalls of analogical reasoning. Nowadays it is usual to recognise that the 'five senses' have many differences and distinctions. Indeed, some physiologists advocate as many as twenty to thirty distinct sensations and classify them on the basis of stimulus, sensation, sensory mechanism and type of response (Moncrieff 1951: 54). Of course, where the similarities in sensory mechanism override the differences—namely in the case of taste and smell—Prout's reduction holds good. However, the investigation of these chemical senses, as of the elements, their atoms and their nuclei, has been by experimentation and not solely by speculative analogical thought.

In the present context of physiology Prout (1810a:16) was not concerned with speculations about the decomposition of the elements; so, dismissing them, he returns to his attempt to show that sensation is really only of one kind. Various analogies between the different kinds of sensation are next considered. Although roughness and taste may seem to be very different they are analogous in the following respects: both may vary in intensity and quality within certain fixed (though admittedly undetermined) limits. 'Perhaps', Prout (1810a: 17) suggests in a fit of Pythagoreanism, 'the varieties of all the modifications of sensation are caused by varieties in the particles of matter producing them whose ratios with respect to each other are contained within 1 and $\frac{1}{2}$ only, that is to say that they are the extremes of the ratios of the particles of matter producing such varieties'. The mathematical reasoning, which is not carefully explained, is taken from the analogy of sound and it leads Prout to the suggestion that there may be at least three 'primitive tastes' which correspond to the three principal sounds (the octave or fundamental, the third, and the fifth), and the three principal colours, red, yellow and blue. The few other states experienced might then be supposed to be caused by particles whose ratios with respect to each other were intermediate between 1:1, 4:5 and 2:3. Wisely, Prout does not attempt to identify the three principal tastes beyond making the suggestion that they may be acid, alkaline and bitter. Intriguingly, modern physiologists agree with Prout, but on experimental grounds, that most tastes are blendings of a few primary tastes of which the four usually recognised are sweet, sour, salt and bitter (Moncrieff 1951: 131; Mitchell 1956: 185).

Similar analogies could be worked out for tasting and smelling; but it is Prout's views on sound which best reveal his belief in underlying mathematical harmonies in nature. It will be recalled that he believes sound to be conducted by vibrations of a Newtonian ether, not the air, and that hearing is produced by the striking of successive ether atoms against the auditory nerves. The sensation of hearing varies with the size, shape and nature of the bodies which produce the sounds; and irregular rough surfaces produce irregular rough sounds because the aggregate particles of ether vary in size from the very large to the very small. If all the particles are of one size then the sensation produced will be smooth and uniform and vary only in gravity (i.e. pitch). By the analogy of light and the musical octave, Prout (1810a: 19) is led to postulate the existence of seven primitive sounds.

> When one undula is performed in the same time as any two others the *octave* is produced by such two undulae. Now the times of these undulae are as the sizes, &c of their causes and hence there are seven points between the two extremes of the octave including the first which are more marked than the rest on account of the simplicity of their ratios &c and which thus constitute seven primitive sounds gradually rising above each other in intensity of which the principle, third and fifth are the chief.

Loudness is dependent on the number of vibrations which act on the ear in the same time and on the force with which they strike the auditory nerve in the ear.

Following an analogy drawn by Newton (1952: 125) in his *Opticks*, it seemed likely that the seven primitive sounds corresponded to the seven primitive colours. Newton had described how he had measured the bandwidths of the spectral colours and found that if the base of the spectrum, GM, was extended to X, so that $GM = MX = \frac{1}{2}$, with $GX = 1$, then

$$\frac{G\lambda}{GX} = \frac{8}{9} \text{ (violet)} \qquad \frac{G\nu}{GX} = \frac{5}{6} \text{ (indigo)} \qquad \frac{G\eta}{GX} = \frac{3}{4} \text{ (blue)}$$

$$\frac{G\varepsilon}{GX} = \frac{2}{3} \text{ (green)} \qquad \frac{G\gamma}{GX} = \frac{3}{5} \text{ (yellow)} \qquad \frac{G\alpha}{GX} = \frac{9}{16} \text{ (orange)}$$

$$\frac{GM}{GX} = \frac{1}{2}.$$

Accordingly, Newton suggested, the boundaries of the different colours were 'in proportion to one another, as the numbers 1, 8/9, 5/6, 3/4, 2/3, 3/5, 9/16, 1/2, and so ... represent the Chords of the Key, and of a Tone, a third Minor, a fourth, a fifth, a sixth Major, a seventh and an eighth above that Key'. Newton was particularly struck by this analogy between the spectrum and the musical scale since he mentioned it on several occasions. Prout must have read Newton's clearer account of the analogy in the posthumously published *Lectionis Opticae* (1744) since he referred to the fact, not mentioned in the *Opticks*, that the breadths of the orange and indigo bands were approximately equal to only a half of the other colours (Newton 1744:ii, 245; 1959:376). In his *Optical Lectures* Newton had suggested that indigo and orange were analogous to the semitones, and the other five colours to the tones, of a musical scale. As Houstoun (1923:12) and others have pointed out, the resultant symmetrical scale is the Dorian mode on D; DEFGABCd.

Prout was obviously captivated by this analogy and suggested that such a scale might be used by composers with advantage. 'Indeed, if my memory does not deceive me I have read of strains having been occasionally introduced into their composition in this mode by some celebrated masters with very fine effect'. The analogy has, of course, been exploited by twentieth-century composers such as Alexander Scriabin who directed that a 'keyboard of lights' was to accompany his *Prometheus—Poem of Fire* (1911; Scholes 1970:202). Red, yellow and blue were the colours which stood out in the spectrum (the modern theory of colour sensation substitutes green for yellow) and they corresponded with the fundamental, third and fifth of the octave. The sensations of colour arose from the mixture of these three primitive colours. In view of the continual interest in blind people who claim to be able to distinguish colour by touch, it is notable that Prout quotes this phenomenon as a possibility because colour, like touch, is dependent on the state of the superficies of aggregated matter.

In fact, of course, both Newton and Prout were completely deceived. There is no correlation between the light spectrum and musical scales. Newton was misled by the character of his glass prism since the widths of spectral bands depend upon its characteristics. In any case, Newton's measurements of these widths were so inaccurate as to suggest that he had already conceived the analogy between light and sound before the measurements were made. This is one possible reason why Newton abandoned any attempt to overcome chromatic aberration (Houstoun 1923:16). Although the methods for removing such aberration had been known for over sixty years, Prout clearly did not appreciate the experimental fallacy behind Newton's analogy. If he had measured the spectral bandwidths for himself, Prout would have been unable to find any analogy with musical intervals. Instead, committed to the discovery of analogies between different sensations, he was glad to accept, uncritically, the support of the illustrious Newton.

To complete the cycle, Prout finally exploits analogies between audition and vision just considered and gustation and olfaction. Perhaps, he speculated, the three principal sounds and colours correspond to three principal tastes and smells. It is quite fortuitous that the modern theory of gustation and olfaction actually supports such reductions (Moncrieff 1951: 393; Starling and Evans 1968: 1406).

What advantages were there in Prout's curious analysis of sensation? Would it not have been just as reasonable to have treated the five sensations as distinct phenomena? Prout believed that such a five-fold analysis would not be as simple; nor would it suggest 'the coincidence which in this respect it points out between this and the rest of [God's] works'. Economy of thought, simplicity and design were also supported by the view of the unity of matter and life which Harris (1775:88) had announced previously.

> And thus we may affirm that these three, that is to say, EXTENSION, FIGURE, and ORGANIZATION, are the three ORIGINAL FORMS to BODY PHYSICAL or NATURAL, *Figure* having respect to its *External*; *Organization* to its *Internal*; and *Extension being common* both to one and to the other. 'Tis more than probable that from the Variation in these *universal*, and, as I may say, *Primary Forms*, arise most of those *secondary* Forms usually called QUALITIES SENSIBLE, because they are the *proper* Objects of our several Sensations. Such are Roughness and Smoothness, Hardness and Softness, the tribes of Colours and Odours, not to mention those Powers of Character more *subtle*, the Powers Electric, Magnetic, Medicinal, &c.

One difficulty remained. How could the same kind of matter produce roughness to the touch, or taste to the tongue? Prout (1810a: 22) was unable to give a specific answer to this although he pointed out that it was clear from the behaviour of chemical elements that different arrangements and combinations could produce strikingly different properties. Why should the same not be true of sensation?

> The same variety of matter in innumerable instances is capable of producing by admixture or combination with another variety of matter in different proportions &c. only modifications equally striking in their apparent forms & qualities as for example oxygen and nitrogen.

As we shall see, this is 'Prout's hypothesis' which was to be placed upon an experimental basis by its author, in an atomic context, five years later. It is on this high point that Prout concludes his essay.

De Facultate Sentiendi is an extraordinary and excellent example of the power and pitfalls of analogical reasoning which is unguided by experiment. Humphry Davy (1816) was to write only a few years later that 'the substitution of analogy for fact is the bane of chemical philosophy, the legitimate use of analogy is to connect facts together, and to guide to new experiments'. If Prout achieved a number of intriguing generalisations about the nature and etiology of sensation in the essay, it was at the cost of going far beyond the limited empirical evidence and over-emphasising superficial or accidental similarities and missing the significant differences between sensations. Although Prout was never to develop these ideas on the chemical and unitary basis of sensation, the quest for a simple chemical picture of the world was to remain with him throughout his career.

With these remarks in mind it is perhaps significant that the notes on taste, smell and flavour which closed the essay are those which have best stood the course of time; for the basis of his discussion here is firmly empirical. These notes were also important to him professionally since they formed the basis of his first publication. Until Prout argued correctly that taste and flavour were different sensations, the two were always confused, physiologists believing that flavour was the principle which excited taste.

> Taste is that modification of sensation which is caused by the contact of certain substances soluble in water or saliva with the tongue, the nostrils being at the same time closed and the tongue not being in contact with any other part of the mouth [Prout 1810a: 24; 1812a: 457]

For the first time in the essay Prout mentions experiments which support this definition of gustation. If substances like nutmeg were placed upon the tongue and the nostrils closed or plugged, then only a slight pungency was experienced. This was in fact the empirical justification for Prout's contention expressed in the main essay that the number of tastes was limited to acid, alkaline, bitter 'and perhaps one or two more'.

Since substances which excited taste usually also excited sensations when placed in contact with parts of the body stripped of the cutis, Prout concluded, somewhat shakily, that all substances which excited taste were stimulants. This, of course, fitted in with Prout's claim that taste was similar to touch, and yet the most limited and imperfect of all sensations. What is usually taken for taste, says Prout, is flavour, which is really a combination of taste and smell. People with colds who lose their sense of smell also lose their sense of 'taste', i.e. of flavour; and people born without a tongue could legitimately claim to 'taste'.

> *Smell* is that modification of compound sensation which is excited by various states of matter either in an aeriform state or in a state of extremely fine mechanical division when these are drawn in the air through the nose [Prout 1812a: 458]

The sensory mechanism of the nose, like the tongue, was chemical or galvanic. Flavour was another modification of sensation produced by the union of taste and smell. 'Substances in general have the strongest flavor that are volatazable or partly soluble in air as well as water'. In his published paper Prout (1812a: 460) gave flavour the rigorous definition which has since passed into scientific literature.

> *Flavor* is that sensation which is produced when substances under certain circumstances are introduced into the mouth, *the nostrils being at the same time open.*

Olfaction was independent of gustation, but not of flavour. On the other hand, olfaction could be influenced by strong flavours, as when vinegar is held in the mouth and masks the odour of ammonia held to the nose.

These remarks of taste, smell and flavour were published anonymously by Prout in 1812. Later they were noticed by his friend, John Elliotson, who had been a student with him at Edinburgh, and published in Elliotson's translation of Blumenbach's *Institutions of Physiology*. In this way Prout's analysis passed into the general literature of the physiology of sensation, though curiously his name was never attached to it (Brock 1967; Starling and Evans 1968: 1410).

Digestion

One of the more interesting papers to be published in the short-lived *Annals of Medicine and Surgery*, which was probably edited by Prout and Elliotson (Brock 1964), was Prout's *Inquiry into the Origin and Properties of the Blood*. Only three parts of this review (much of it based upon his London animal chemistry lectures) were published in 1816 and because of the journal's limited circulation Prout had the incomplete paper republished in 1819 in a slightly modified and abbreviated form in Thomson's *Annals of Philosophy*, from where it gained French and German translations. A missing section, which had entailed a long investigation of the development of a chicken's egg, was only finally presented, as an independent paper to the Royal Society, in 1822.

Prout's announcement of the discovery of hydrochloric acid in animal gastric juice followed in the autumn of 1823. But Prout had not suddenly started to work on the chemistry of digestion in 1823 for, as the paper on sanguification shows clearly, he had been involved in the problems of digestion as far back as 1816 when he had argued that 'blood begins to be formed, or developed from the food, in all its parts from the first moment of its entrance into the duodenum, or even, perhaps, from the first moment of digestion, and that it gradually becomes more perfect as it passes through the different stages to which it is subjected, till its formation be completed in the sanguiferous tubes, when it represents an aqueous solution of the principal textures and other parts of the animal body to which it belongs' (Prout 1816a: 20). At this time he divided the process of blood formation into four stages: digestion (confined to the stomach), chymification (confined to the duodenum), chylification (confined to the lacteals) and sanguification proper (confined to the blood vessels). This suggests that his concern to provide a chemical natural history of blood formation led him to the particular study of digestion as part of this larger programme of physiological research. In fact the programme probably began during the lecture period of 1814, for he appears to have received encouragement to do this from both Alexander Marcet and Astley Cooper (Scudamore 1817: 509; Berzelius 1912–36: i, 123).

Although Prout reviewed each of the four stages of sanguification in turn, here we shall restrict attention to his work on digestion. The acidic contents of a rabbit's stomach were examined some two hours after feeding and shown to contain 'traces of alkaline muriate, with slight traces of an alkaline phosphate and sulphate; also of various earthy salts, as the sulphate, phosphate and carbonate of lime' (Prout 1819a: 13). A similar acidity—a well authenticated observation in the contemporary literature (Kasich 1946)—and analysis were found in the gastric fluids of a pigeon, tench and mackerel. Since the heterogeneous nature of the stomach fluids had caused so much confusion in the past, Prout proposed to divide and identify the potential contents into saliva, the mucous coat and exhalents of the stomach, and the actual gastric juice whose identity was 'unknown, it never having been attained in a separate state'. Yet it was clearly 'some volatile acid from its effects on litmus paper. ... I considered it in the pigeon as carbonic. There appears, however, to be occasionally another acid which is of a much more permanent nature, and it is probably the phosphoric acid' (Prout 1819a: 272). It is clear, therefore, that at this period, 1816 to 1819, Prout was a long way from the identification of the gastric juice as hydrochloric acid; in fact, by 1820, he had come to accept Berzelius's opinion that the active principle was lactic acid (Berzelius 1813b: 207).

In his study of the 'evolution' of the blood, Prout (1819a: 13) analysed the chymes—that is the contents of the small intestines—of several animals, including dogs, rabbits, an ox, tench, mackerel and a pigeon fed on vegetable diets in 'order to ascertain if the chyme exhibited any traces of the albuminous contents of the blood' which the food contents had not. He used Berzelius's test for albumin, which involved the addition of potassium ferrocyanide to an acetic acid solution of the substance; a white precipitate indicated the presence of albumin. Despite the changes which obviously took place in the chyme after it had passed through the duodenum and mixed with the bile and pancreatic juices, Prout could not detect any albumin. In view of Beccari's demonstration in 1742 that vegetable substances contained albumin, Prout's experimental results are distinctly odd, and they puzzled later investigators like Tiedemann and Gmelin.

It took the skill of Claude Bernard (Holmes 1974) to notice the analogy between the changes which took place in the duodenal chyme and the saponification of fats in the laboratory, and to ask what was the chemical cause of this change. Yet Prout (1819a: 273), like Spallanzani before him, believed that the changes were purely and simply chemical, even though they were ultimately under the control of vital forces. They were, therefore, reproducible under laboratory conditions.

> I tried to produce these changes out of the body, and with this view mixed a portion of the fluid obtained from the contents of the stomach of the rabbit, ... with a portion of the bile of the same animal. A distinct precipitation took place, and the mixture became neutral; but although I thought that the resulting fluid was more of an albuminous nature, yet the formation of a perfect albuminous principle was doubtful; probably the presence of the pancreatic juice was necessary to complete the formation of this principle.

When Prout continued the analyses of the contents of the alimentary tract through the large intestine, caecum and rectum, he found that the 'albuminous nature' vanished altogether.

The experiments on 'the evolution, &c., of the blood of the chick *in ovo*' were not completed in time for inclusion in the paper of 1816; nor, because of his work on urine chemistry, was he in a position to publish any results when the sanguification paper was republished in 1819. The chick embryo investigation had meanwhile broadened in scope, and in 1822 it was presented to the Royal Society as 'experiments on the changes which take place in the fixed principles of the egg during incubation' (Prout 1822b). This was one of the first examples of chemical embryology to be published—a field of research which has only produced significant results in this century.

Prout found that an incubating egg lost about a sixth part of its weight by evaporation of water though its shell; this loss was some eight times greater than from an egg placed under non-incubatory conditions. In eggs examined analytically over an incubation period of three weeks, Prout found that an exchange of chemical substances took place between the yolk and a part of the white. Initially the yolk (which was very difficult to analyse since combustion brought about the oxidation of phosphorus to phosphoric acid, which inhibited complete combustion) lost a portion of its oily matter to the white causing it to curd like milk. Reciprocally, the white lost portions of its watery and saline parts to the yolk which increased in size. As the incubation proceeded, the watery and saline parts migrated back to the white from the yolk, which consequently shrank in size. In particular, Prout detected a loss of phosphorus from the yolk to the white where it formed the calcium phosphate essential for the growth of the chick's bones. However, the overall amount of phosphorus in the embryo remained constant. On the other hand, where had the calcium come from, for according to Prout's analyses there was a deficiency of calcium in the new-laid egg. The only possibilities were the shell, or from 'the transmutation of other principles'.

Prout claimed that it would be impossible for an analyst to demonstrate that the shell provided the lime (which, of course, it does) because of the wide variations in shell weight and the impossibility of analysing the shell before incubation. Furthermore, there appeared to be no vascular connections between the shell and the membrane putananis which inscribed it; how then could calcium pass from the shell to the membrane? Hence, although in 1816 Prout had strongly denied the possibility of a creation of elements by animal organisms, he now proposed for serious consideration the possibility of a synthesis of the 'element' calcium. He carefully guarded himself against ridicule:

> I by no means wish to be understood to assert, that the [calcium] earth is *not* derived from the shell; because, in this case, the only alternative left to me is to assert that it is formed by transmutation from other matter; an assertion, which I confess myself not bold enough to make in the present state of our knowledge, however strongly I may be inclined to believe that, within certain limits, this power is to be ranked among the capabilities of the vital energies (Prout 1822b: 400).

As we shall see, Prout did not believe in the ultimate nature of the so-called elements, so that for him, and several other chemists and vitalist physiologists like Daubeny (1850:374) and Elliotson (1828:399) such organic transmutations were legitimate speculations. Sir Gilbert Blane (1825:70) took Prout to mean that elements were created in the body by a vital agency, but this was later refuted by Alfred Taylor (1849). Prout's anonymous (1851) Edinburgh obituarist saw a moral in this episode:

> The explanation which, in the light of twenty-six years chemical analysis has been able to furnish, is most serviceable in impressing the lesson, that, in circumstances in which we are unable to explain all phenomena in living bodies, it is most prudent to wait and suspend judgment until the accumulation of further facts ... comes to give the elucidation which was previously wanted.

Berzelius (1825), who represented the multiple-elements chemists and physiologists, held no brief for transmutations. He therefore studiously avoided mentioning Prout's speculation in his fullsome review of Prout's paper.

> I believe that the purely chemical part of this research has been one of the most important inquiries in animal chemistry during the past year; it should not be ignored.

Indeed, it was a pioneering foray into chemical embryology; but despite Berzelius's words the paper was destined to be forgotten in the advent of superior analytical techniques and the development of colloid chemistry.

By 1822 Prout was so completely in command of his material that (having just completed his treatise of urinary diseases) he was able to plan a book on digestion to be called *Observations on the Functions of the Digestive Organs, especially those of the Stomach and Liver; with Practical Remarks on the Treatment of some of the Diseases to which these Organs are Liable*. A prospectus published in the *Annals of Philosophy* in February 1823 announced that the work would:

> comprise the results of an experimental inquiry into the nature of some of the more important chemical changes which take place during the digestion and assimilation of the food. The practical remarks will principally relate to the proper adjustment and use of remedies, and to the pernicious effects liable to be produced in delicate habits by the constant operations of variously slowly acting causes, especially impure or hard waters: illustrated by analyses of the principal waters in common use in the metropolis and its vicinity.

Although this book actually went to press, it was never published; on the other hand, much of its intended contents found their way into the *Bridgewater Treatise* of 1834 and the enlarged edition of his urinary textbook in 1840. John Elliotson (1828:310) and Herbert Mayo (1833:152) also had access to it and quoted extensively from it. The explanation given later by Prout for its non-appearance was that his discovery of hydrochloric acid in the gastric juice during the autumn of 1823 so thoroughly upset his assembled materials and ideas on the subject of stomach digestion that he felt obliged to withdraw the book. That this drastic action was necessary appears surprising in view of the small extent to which hydrochloric acid figured in his later theory of digestion. That there were other factors was clearly suggested by the anonymous (1851) writer of the *Edinburgh Medical and Surgical Journal* obituary notice of Prout.

The first edition of Prout's urine textbook had proved successful. It had the distinction of being one of the first English medical treatises which merited translation into French. The publisher's stock was exhausted and a new edition demanded. As we have seen, this second edition, with several new features, appeared in June 1825. Consequently, as a direct result of the publication of the urine textbook and his growing reputation as a urinary specialist, Prout's medical practice increased considerably and the time for his own research and writing was drastically decreased. When Berzelius (1912–36: i, 232) politely inquired of his friend Marcet for news of London chemists, including Prout, Marcet replied significantly: 'Prout is well, but unfortunately he is a doctor, and his researches are too much interrupted'. Such little time as he did have for research at the end of 1823 seems in any case to have involved the analyses of wines for an old Edinburgh friend, Alexander Henderson. The latter's *History of Wines Ancient and Modern* (1824) acknowledges Prout's help in several places and includes Prout's analyses of eighteen wines in an appendix.

These reasons, coupled with the hydrochloric acid discovery, would undoubtedly explain why the book was abandoned. And if the work was initially only postponed, a decisive factor for abandoning it completely would have been the publication by the German physiologist and chemist, Tiedemann and Gmelin (1828) of their fine study of digestion. In this connection a further reason will become clear later.

Prout's discovery of hydrochloric acid in gastric juice was announced to the Royal Society on 11 December 1823. Now recognised as 'a classic of scientific reasoning', the published account in the *Philosophical Transactions* is extremely terse; so much so that although most chemists and physiologists appear to have agreed with Prout's identification—especially after the spectacular confirmation from Beaumont (1833) in America—there remained a few who were prepared to argue the validity of Prout's analytical deductions. Immediate challenges came from Leuret and Lassaigne (1825) in France, who claimed to have found free lactic acid in the stomach, and from Tiedemann and Gmelin (1828), who had detected free acetic, butyric and hydrochloric acids. Both groups of investigators were engaged on the problems of digestion at the same time as Prout, and the resulting controversy (Mani 1956; Holmes 1974: 149) enables a more detailed account of Prout's discovery to be made than is possible from the short account he gave to the Royal Society.

It will be recalled that Prout's tentative view between 1816 and 1819 was that the active acid of the gastric juice was the phosphoric, mixed perhaps with some carbonic acid; but that by 1820, when he actually identified free hydrochloric acid in the gastric fluid of a dyspeptic patient, he had been conditioned by Berzelius into believing that the active principle was lactic acid. Any hydrochloric acid which he found was dismissed by him as pathological, or as due to bad analysis. More than likely he intended to present in his book on digestion some sort of analytical proof that the acid was lactic and no other. Consequently, he wrote later, my 'inquiry was conducted in a much more rigorous and elaborate manner than it probably otherwise would have been'. Eventually, 'after a series of the most complete evidence that perhaps was ever brought to bear on a chemical point, I was obliged to conclude, in opposition to my preconceived notion, that the acid was the muriatic and no other' (Prout 1828: 120). (Muriatic acid was the early nineteenth-century term for hydrochloric acid.)

Now the misunderstanding between Prout, Tiedemann and Gmelin, and to a lesser extent with Leuret and Lassaigne, arose because Prout did not publish any description of his preliminary experiments, or mention his previous commitment to lactic acid; instead, in the manner of a mathematician offering a proof of a theorem, he offered a demonstration to his readers that the gastric acid was hydrochloric acid and a value for the quantity of free acid in the stomach. Prout's methodology here was sound.

> The mere determination of the existence of a principle in any compound, without its quantity be at the same time ascertained, is often unsatisfactory; at least the determination of the latter point corroborates the former in no small degree; but before the quantity of a substance can be ascertained, it must be obtained *per se*, or in some well-determined state of combination, circumstances necessarily implying a much more complete and satisfactory investigation than that by *mere tests* alone [Prout 1828: 120].

He had spent much time devising a method for the characterisation of gastric juice which would satisfy this criterion, and he had been able to describe his results to the Royal Society economically, without detailing all his many starts and failures. For example, he had found that distillation of gastric juice (a method used by Tiedemann and

Gmelin) only proved the presence of hydrochloric acid and could not be used to determine its quantity.

For his demonstration, Prout (1824a) washed the contents of a rabbit's stomach in distilled water. After filtration he divided the resultant clear liquid into four parts. The first portion (1) was used to determine the quantity of hydrochloric acid (i.e. chloride) present in the fixed alkaline chlorides, by precipitation with silver nitrate. Since the solution was previously evaporated to dryness, the volatile ammonium chloride and free hydrogen chloride had to be calculated separately as follows. A second portion (2) was supersaturated with potash to neutralise the free hydrochloric acid, and the total quantity of chlorides, including ammonium chloride, was found by precipitating with silver nitrate. In their treatise on digestion, Leuret and Lassaigne (1825: 112) voiced disapproval of this step: neutralisation with excess potassium hydroxide, they claimed, would have led to many side products and pure silver chloride would not have been precipitated. Hence Prout's calculation for free hydrochloric acid was invalid. In reply Prout pointed out that 'the merest tyro' would have obviated this defect, and he retorted sarcastically 'I wish these gentlemen to know, what every *chemist* might have taken for granted when it was stated that the experiments "were made in the *usual manner*", that the excess of potash was always supersaturated with nitric acid before the nitrate of silver was employed'.

The third portion (3) was used to determine directly the quantity of free acid by volumetric analysis using a solution of potash of known strength. The combination of these results (1, 2, 3) determined the amount of chlorine that was combined with ammonia $(2 - (1 + 3))$. This last result was independently checked by evaporating to dryness the neutralised third solution in order to volatalise the ammonium chloride. The amount of chloride combined with fixed alkalies was then determined with silver nitrate (3A). (Quantity of ammonium chloride equals portion 2 less 3A.) In a later comment Prout (1828) stated that he had omitted to mention at this point in his original paper that the solute remained neutral after the evaporation. This, he argued, could not have been the case if any combustible acid had also been present in the gastric juice, for then the carbon dioxide released would have produced some acidity. The purpose of the fourth portion of gastric fluid was in fact to test for the presence of other mineral acids, particularly sulphuric or phosphoric acids; neither was detected.

Prout's analyses demonstrated that there was a very considerable quantity of unsaturated free hydrochloric acid in the stomachs of rabbits, horses, calves and dogs, during digestion. Examination of the fluids ejected by dyspeptic humans confirmed its presence in man. No ammonium chloride was found in human fluid—except in the case of one dyspeptic who found solace in drinking ammonia! The presence of such free hydrochloric acid was rapidly confirmed by J G Children (1824), the Secretary of the Royal Society, from the observation of a dyspeptic and sceptical colleague. It was also confirmed independently by Tiedemann and Gmelin in February 1824.

However, the Germans, who together were more than Prout's equal in the field of the chemistry of digestion, remained unconvinced that hydrochloric acid was the only significant acid present in gastric juice. In 1823 the French Academy of Sciences had proposed as a topic for an essay competition the nature of digestion in different classes of animals. Not surprisingly Prout prepared to enter the competition with his book on digestion, which he accordingly withdrew from the press. Here, then, is yet another reason why the work remained unpublished. In July 1824 he told Walter Adam that 'I shall give up all idea of writing for the French Institute. The award I understand does not take place till Midsummer next, which is too long for me to wait, and perhaps be disappointed after all' (Prout 1824b). Ironically, this turned out to be a wise decision for none of the candidates entirely satisfied the Academy's referees, though honourable mention was made of Leuret and Lassaigne, and Tiedemann and Gmelin, who 'have made a great number of experiments, and have attained remarkable results. For this reason, and in consideration of the expensive nature of the researches in which they engaged, the Academy have ajudged to each [party] the sum of 1500 francs'. Leuret and Lassaigne accepted, but the Germans were offended by the Academy's offer which carried the implication that their work was no better than that of the French team; they therefore published their work independently in German (Holmes 1974:149). The whole matter assumed the proportions of an international academic scandal—especially after the publication of the respective researches revealed their comparative merits for all to see. Prout (1828) himself was in no doubt who should have won.

> I have a high opinion of MM. Tiedemann and Gmelin's volume. Having gone over most of the ground traversed by these gentlemen, I am well aware of the labour and difficulty of the march; and though we may differ in some minor particulars, which is not to be wondered at, I am satisfied, as far as we go together, with the general accuracy of their observations. With respect to MM. Leuret and Lassaigne's book, I am sorry that I cannot express the same sentiments; indeed as a work it does not appear to me to be at all comparable with that of the German philosophers.

The German investigators had, in fact, admonished Prout for appearing to deny the presence of any other acids in gastric juice—

they had detected both acetic and butyric acids in the stomach, as well as hydrochloric acid. But this was due to a misunderstanding caused, as we have seen, by the terseness of Prout's original paper. In 1828 Prout readily conceded to Tiedemann and Gmelin that other organic acids were sometimes present, 'though I confess I believed, and still do, that the muriatic acid occurs naturally more frequently, and in greater abundance in that organ than any other acid'. Such acids, he reasoned, were derived from the food in the stomach, rather than from the stomach itself; alternatively, their presence was an indication of the derangement of the stomach and of the process of assimilation.

Strong support for Prout's discovery came from the observations of the American doctor, William Beaumont (1833). In 1835, Braconnot (1836) confirmed the presence of hydrochloric acid and the absence of lactic acid in the gastric juice of a dog. Nevertheless, a persistent group of sceptics, including the young Claude Bernard, continued to suggest that the only genuine acid was the lactic, while the hydrochloric acid was only produced by interaction with food chlorides— the very process by which Prout believed lactic and acetic acids were formed (Holmes 1974: 223). Sometimes these critics showed a surprising lack of analytical skill. One obstinant opponent was Robert D Thomson (1839), a lecturer at St Thomas's Hospital in London and a nephew of Thomas Thomson. He attacked Prout, and Tiedemann and Gmelin, before the British Association in 1839 and in several lengthy articles. Mildly, Prout noticed in his clinical textbook that Thomson's methods were not to be relied upon. Kasich (1946) has noted the way in which contemporary textbooks were usually quite undogmatic; if they did not explicitly deny the presence of hydrochloric acid, or deny that it was the active acid, they always listed it with other organic acids. Indeed, J W Griffith, the editor of the posthumous edition of Prout's *Bridgewater Treatise*, was to claim lactic acid as the active ingredient of gastric juice (Prout 1855: 345). However, following the publication of Bidder and Schmidt (1852), few physiologists doubted that hydrochloric acid, which was naturally secreted by the glandular cells of the gastric mucosa, was the only free acid.

Nevertheless, it seems probable that even Prout deferred to the criticism of 'over-generalisation' with which his study had been greeted in some quarters, for to a large extent he played down the presence of hydrochloric acid in his later writings on digestion. Although he had said (Prout 1828:122) that his discovery 'was one of those leading facts that open up an entire new field of inquiry', he usually considered chlorine as the active principle of stomach digestion, even after the isolation of pepsin by Schwann (1836). Chlorine and hydrochloric acid were, he suggested, produced from the sodium chloride of the blood by galvanic action. The basic oxide released produced the alkalinity of the blood, and after its excretion from the bile, reformed salt in the duodenum.

> We have in the principal digestive organs, a kind of galvanic apparatus, of which the mucous membrane of the stomach, and perhaps that of the intestinal canal generally, may be considered as the acid or positive pole; while the hepatic system may, on the same view, be considered as the alkaline or negative pole (Prout 1834a: 496; 1848: 470).

The Copley Medal winning paper of 1827, *On the Ultimate Analysis of Simple Alimentary Substances*, was the last of Prout's papers on organic analysis. At the same time, it was an introduction to the theory of digestion which he explored at greater length in the *Bridgewater Treatise* in 1834. The paper was supposed to be the first of three essays in which he would in turn discuss the composition and assimilation of the three foodstuffs which he classified as the saccharinous (our carbohydrates), the oleaginous or oily (fats) and the albuminous (proteins). The nutritional origin of this classification was to have been explained in his unpublished book on digestion. Prout's explanation was, however, fortunately quoted by John Elliotson (1828).

> Observing that milk, the only article actually furnished and intended by nature as food, was essentially composed of three ingredients, viz. saccharine, oily, and curdy, or albuminous matter, I was by degrees led to the conclusion that all the alimentary matters employed by man and the more perfect animals, might, in fact, be reduced to the same three general heads; hence I determined to submit them to a rigorous examination in the first place, and ascertain, if possible, their general relations and analogies.

Man then tried to copy Nature's great aliment in his cookery.

> He dissatisfied with the productions spontaneously furnished by nature, culls from every source, and, by the power of his reason, or, rather his instinct, forms in every possible manner, and under every disguise, the same great alimentary compound. This, after all his cooking and art, how much soever he may be inclined to disbelieve it, is the sole object of his labour, and the more nearly his results approach to this, the more nearly they approach perfection.

It was for this reason that man had instinctively added oil or butter to bread, a farinaceous substance deficient in fat, or fattened animals so as to combine the three principles on one plate.

> Even in the utmost refinements of his luxury and in his choisest delicacies, the same great principle is attended to, and his sugar and

flour, his eggs and butter, in all their various forms and combinations, are nothing more nor less than disguised imitations of the great alimentary prototype *milk*, as presented to him by nature [Elliotson 1828: 310; Prout 1834a: 477].

In an attempt to explain the organisation of aliments into the animal body in his 1827 paper, Prout borrowed an idea from the discovery by John Herschel (1824) of the effects of minute portions of 'impurity' on the electrical states of substances. Prout suggested that in order for a substance, such as a sugar, to pass from a crystalline inorganic state to an amorphous organic state, the presence and admixture of 'foreign or extraneous parts' (including, for this purpose, water) was absolutely necessary. This conception probably dates from Prout's earliest work on the sugars in 1817. These foreign bodies, which could be elements other than the traditional organic elements of carbon, hydrogen, oxygen and nitrogen, or simply the elements of water, were not present in definite proportions, but equally diffused through the substance which otherwise retained its definite composition. Any substance so affected, or 'merorganised' (the term 'protoorganised' appeared in the Royal Society abstract (Prout 1815–40: 324)), not only lost its power of crystallisation, but it also completely altered its chemical properties; in the better-known language of Berzelius, it had isomerised. For example, starch had to be considered as merorganised sugar; their analyses were very similar, but their properties and forms were quite different. In fact, starch is a chain polymer of repeated glucose units, and in this sense it can be viewed as glucose or sucrose molecules held together by the elements of water. Prout noticed the ease with which merorganised substances lost any water not essential to their unit of composition when heated below 100°C. However, although Charles Daubeny (1831: 79, 132) thought that merorganisation was an attractive way of explaining certain kinds of isomerism, no other chemist or physiologist adopted the word or the concept, and after 1831 Prout abandoned the word, though not the homeopathic concept.

Although the 'merorganised lucubrations' (Anon 1832) essentially involved the recognition of both the necessity for trace elements in the chemistry of organisation and a realisation of the tremendous significance of water in the economy of living systems, Prout fell a victim to the complication, or unwise depiction of these substances as 'extraneous' or 'foreign'. Other chemists came gradually to recognise their importance in animal economy; but they also saw that it was far simpler not to view them as 'impurities' but as substances which entered into composition in small, but nevertheless definite, proportions.

The Copley paper *On Ultimate Analysis* effectively linked together for the first time Prout's preoccupations with accurate analysis and animal chemistry, and (in view of the suppressed book) anticipated the fuller public expression of his ideas on digestion which he was to publish in 1834. Apart from the speculation concerning merorganised substances, this paper was firmly grounded in experiment and gave the chemical world details of a new method of analysis, a new classification of nutrients and accurate analyses of many sugars. Prout had obviously come a long way from the extravagant undergraduate analogies of *De Facultate Sentiendi*. Since then he had established himself as an expert on urinary and digestive diseases, as an accurate and very conscientious analyst, and as a chemist who was especially interested in physiology. His two papers on respiration, though wanting in interpretation, had been pioneering attempts to achieve precision in the subjection of the human body to quantitative analysis. The difficulties which attended this, and the fact that his work was done only on a limited scale and unrelated by him to digestion and excretion, diminished the value of his work which in any case could not be properly understood until the intricacies of cellular chemistry began to be unravelled towards the end of the century. Again, his work on urine and digestion between 1814 and 1827 provided him with a firm experimental foundation upon which to erect a theory of metabolism. His research on the egg had been an entirely original study of the chemical, as opposed to the anatomical, changes which took place during incubation. Such a pioneering attempt to apply chemistry to the living body was filled with difficulties of interpretation that could not be properly made until after the development of physical chemistry. Finally, Prout's identification of hydrochloric acid in the gastric juice was a sensational discovery, made with great analytical skill, and one which solved a very old problem.

Yet throughout these experimental studies in chemical physiology Prout was not content only to publish practical results. Speculative hints of a daring and very general kind were continually dropped by him: the sun and the cycle of the seasons directly influenced the rhythm of living processes; elementary analyses revealed a numerological pattern which might be a key to the understanding of the mutual relationships and transformations in the living body; the possibility that living systems could, under exceptional circumstances, synthesise elements; the electrical nature of some of the basic chemical processes performed in the living animal; and, finally, the idea that traces of elements mixed with inorganic or organic substances might be the sole difference between them and organised bodies. How these general ideas fitted in with Prout's unitary and molecular theory of matter will become clear in the next few chapters.

4

Natural Theologian

> To dismember *Meteorology* from *Geology*—the one involving *causes* of which the other presents the *effects*—in order to make it the link between *Chemistry* and *Digestion*, was the work of no ordinary mind; and to separate *Digestion* from *Physiology*—a part from the whole—and again place *Physiology* deprived of the *Hand*, in opposition to the *Physical Nature of Man*, allowed to retain his hands, evinced most uncommon tact in classification. The ingenuity with which Dr. Prout has connected his subjects, does not render their combination a bit the less ridiculous [*Athenaeum* 1834: 349]

Although the design argument for the existence of God from 'the wonder of his works' is to be found in the pre-Christian writings of the Greeks and Romans and has always formed a significant part of Christian apologetics, complementing the argument from revelation, it came to particular prominence during the scientific revolution in the seventeenth century. This justification of the scientific enterprise through 'natural theology' remained a subject of considerable popularity during the eighteenth century and, despite logical criticisms by Hume, it survived in Great Britain into the nineteenth century until the theory of evolution demolished its empirical foundations (Hurlbutt 1965). A steady stream of learned and popular literature on the subject was published during the 100 years after Newton's death in 1727. The very vastness of this literature has suggested to historians such as Brooke (1978) that natural theology fulfilled extrinsic social and mediating functions as much as ones which were intrinsic to theological traditions. In particular, because the argument was doctrinally neutral and imprecise, it could appeal to different and often rival religious traditions, denominations and sects. The literature is epitomised by Archdeacon William Paley's hugely successful *Natural Theology* which appeared in 1802, and whose method of argument became indelibly impressed upon the mind of the young

Charles Darwin (1887: i, 47; ii, 219). Paley, who was committed to the mechanical world picture of the English Newtonian tradition—the Sovereign Order of Nature—reacted strongly against the atheistic 'blind chance' of the French savants and of such English radicals as Erasmus Darwin. His arguments were drawn mainly from examples of organic contrivances, that is, he argued that the various species of animals and plants were examples of designed artifacts. Paley paid little attention to the premises of the argument for design (Gundry 1946) and he virtually ignored the devastating criticism of natural theology which had been published by Hume (Wollheim 1963). This is not surprising, for the Christian religion was the *status quo* of an ordered, decent society and, as Paley recognised, the ghastly alternative appeared to be the blind chance, atheism and anarchy of French materialists and revolutionaries.

There were some twenty editions of Paley's manual before 1820, and a further ten appeared before the end of the century. The book became the Bible of Nature for many a scrupulously religious scientist, especially liberally educated physicians. For example, Charlotte Hall (1861), the devoted wife of the physiologist Marshall Hall, portrays her husband engrossed in Paley while his carriage conveys him upon his medical rounds. It is not a coincidence that four of the eight authors of *Bridgewater Treatises* were practising physicians.

The *Bridgewater Treatises* were sponsored by the will, dated 1825, of the eighth and final Earl of Bridgewater, The Reverend Francis Henry Egerton (1756–1829). The son of a bishop, Egerton was an eccentric clergyman who 'assiduously neglected his parish' for studies of French and Italian literature, family history and natural theology (Gillispie 1951: 209). In his will he bequeathed £8000 to the Royal Society as a payment to the person or persons chosen by its President who could 'write, print and publish, one thousand copies of a work on the Power, Wisdom and Goodness of God, as manifested in the Creation; illustrating such work by all reasonable arguments, as for instance the variety and formation of God's creatures in the animal, vegetable and mineral kingdom; the effect of digestion and thereby of conversion; the construction of the hand of man, and an infinite variety of other arguments' (Prout 1834: Prelim).

These terms placed a considerable burden of responsibility on Davis Gilbert, a patron and personal friend of Sir Humphry Davy, who had succeeded the latter as President of the Royal Society in 1827. Gilbert's responsibility was made more difficult by the fact that during 1829 and 1830 he was the centre of the controversy over the government of the Royal Society between aristocratic and democratic men of science (Todd 1967; MacLeod 1983; Miller 1983). Prout, who was on the side of reform and a more democratic form of Society government, nevertheless favoured the continuation of Gilbert as President. 'My dear Sir', he wrote to Gilbert, a gentleman-amateur of science and a Member of Parliament rather than a practitioner of science,

> I cannot help expressing my regret that you are about to retire from the Royal Society: indeed you have been so shamefully used that I do not wonder at it, but still as I have said, I regret the circumstances very much, particularly on account of the Society itself which I fear will be the sufferer. As to your proposed successor I know nothing; public report represents him as a strong political partizan; as a man of no science, and moreover in such a state of health as to be generally unable during the winter to venture out in the evening. If all this & more to the same effect be true, I cannot conceive an individual worse adapted for the Presidency of the R.S. & I have no hesitation in saying that, on *these grounds*, I as a member, shall most certainly oppose his election. These I believe are the sentiments of a very large proportion of the members, perhaps indeed of most of them, except perhaps the intriguers who, I fear are at the bottom of this, and of all the cabal that has lately existed in the Society & which has so disgusted me that I feel a very strong inclination to retire from it altogether. Still hoping that this great evil will not be inflicted on the Society & that you will long continue its President [Prout 1830].

However, Gilbert's position finally became intolerable and he resigned from the Presidency in November 1830 in favour of the Royal Duke of Sussex. Although Prout, along with eighty other Fellows of the Society, voted for the astronomer, J F W Herschel, in the contested election, the Royal Duke became the Society's president. Gilbert, therefore, did not have a direct hand in the final printing and publication of the *Bridgewater Treatises* between 1833 and 1836; this function was directed by the Society's polymathic Secretary, Peter Mark Roget, who was himself one of the chosen authors. However, before his resignation, Gilbert coordinated the initial search for suitable authors. His personal decision that the legacy should be broken down into eight divisions did not go uncriticised and it became used as yet another source of dissatisfaction with Gilbert's presidency. He had, however, wisely aligned both the Archbishop of Canterbury, William Howley, and the Bishop of London, C J Blomfield, on his side, and invited them to suggest names and even to negotiate terms with writers of their own choice (Enys 1877; Brock 1966). In this manner, Charles Bell was selected for anatomy 'including of course Lord Bridgewater's favorite topic the Human Hand' (Enys 1877: 9); Roget for physiology and comparative anatomy; William Buckland for geology and mineralogy; John Kidd for the adaptation of man's environment to his physical condition; Thomas Chalmers for the adaptation of man's environment to his moral and mental state; William Whewell for astronomy and general physics; and William Kirby for the habits and instincts of animals.

The collection of correspondence connected with the Bridgewater authors that was published privately by Gilbert's nephew, John Enys, later in the century shows that at first Gilbert did not think in terms of a separate chemical treatise on natural theology but one which would include a discussion of the imponderable fluids, light and heat. The name of the evangelical Scots physicist, David Brewster, was proposed and seriously considered; but, prompted by the Bishop of London, by 2 September 1830 Gilbert was instead suggesting a treatise 'Chemistry chiefly in reference to the Etherial or Imponderable Fluids or especially to Light, including Optics with the recent discoveries of Polarization'. Such a book was to be written by either Prout or Brewster. Blomfield (Enys 1877: 20) then advised Gilbert that Prout was a 'first-rate chemist' and that 'I should think would do whatever he undertook with great ability'—though he immediately spoiled the effect by admitting that 'I don't know what sort of *writer* Dr. Prout is; never having seen his publications'. Blomfield concluded his advice by emphasising that the literary merits of Prout and Brewster were an important factor in deciding between them. In fact there was little to choose between their respective literary gifts and scientific experience, and how the decision was ultimately made cannot be known; for with Gilbert's resignation, the task of organisation seems to have passed completely into Roget's hands. It may, however, be conjectured that, since Prout was well known to Roget and Gilbert through his Council membership of the Royal Society, his sympathy for Gilbert in his troubled presidency, and above all, because he was an expert on the subject of digestion which had been specifically mentioned in Bridgewater's bequest, Brewster was passed over in favour of Prout. In any case, Brewster's much publicised anti-Oxbridge feelings would have counted against him. A letter from Blomfield to Gilbert of 18 November 1830 named Kidd and Buckland as his own choice, and Whewell and Chalmers as the choice of the Archbishop of Canterbury. Gilbert (Enys 1877: 24) added to this letter in his own hand the names of Roget, Bell, Kirby and Prout.

During the winter of 1831 scandalous rumours began to appear in the press concerning the bequest in which commissioned authors were spoken of as 'competitors for the legacy' (Anon 1831). Buckland was appalled and appealed to Gilbert for a public report from him concerning the chosen authors. Gilbert obliged with a suitable statement in March 1831. Meanwhile, Roget had difficulty in finding a publisher. John Murray accepted the series, then backed down; Longman refused to publish an initial limited edition of a thousand copies. At last, at a meeting of the nominees in London on 11 October 1832 the final allocations were decided and contracts exchanged with the publisher William Pickering. Lord Bridgewater's desire for professional *a posteriori* demonstration of the existence of a benevolent deity was finally satisfied by the publication of the eight 'strange and deadly' volumes intermittently between 1833 and 1836.

Prout's *Bridgewater Treatise, The Chemistry, Meteorology and the Function of Digestion considered with Reference to Natural Theology*, appeared in the spring of 1834 (prefaced 3 February) and a second edition was found necessary during the same year (prefaced 7 June). Prout was well qualified to write with distinction and originality on the apparently unrelated topics of chemistry and the digestive process; and with distinction, if not originality, on meteorology (Brock 1970). (One of Prout's neologisms in the *Meteorology* was, however, destined to enter the scientific literature: 'convection' for the conduction of heat through air or liquids.) Almost all the reviews were full of praise, although by 1854, when the agnostic physicist John Tyndall (1854) was asked to edit a posthumous edition of Prout's book for the popular Bohn Library series, he commented sarcastically in his private journal:

> I should have thought more highly of Dr. Prout had I not read his book. Certainly if no better Deity than this can be purchased for the eight thousand pounds of the Earl of Bridgewater, it is a dear bargain. It is very evident that Dr. Prout would never have written such a book through the spontaneous promptings of his own spirit; it was written for money, and lacks even common scientific depth, not to speak of religious inspiration.

Tyndall was unfair in his estimate of Prout's scientific competence, for many of Prout's pages had simply dated. Tyndall, of course, was one of the more outspoken Victorian representatives of scientific naturalism (Turner 1981) and his real dislike was for the design argument and the vitalism which pervaded Prout's pages on digestion, Possibly there was a grain of truth in the financial sneer, though this was a common criticism which all the Bridgewater authors had to face. Since 'expense was not regarded by Dr. Prout' (Munk 1878: iii, 111) in acquiring analytical apparatus, and as he had a wife and six children to support, and if the report that he was charitable with regard to patient's fees is not mere 'obituary sentiment', he would have been foolish and extremely high-principled to have rejected the unsolicited legacy. Nor is there any reason to doubt Prout's religious sincerity or belief in the design argument; he rented a pew in the gallery overlooking the pulpit in the fashionable physicians' church of St James in Piccadilly, close to his home in Sackville Street.

In his introduction to the *Chemistry* Prout (1834a) suggested that there were three classes of objects which could be utilised in the argument for design.

1. Those objects regarding which, the reasoning of man coincides with the reasoning evinced by the Creator.

For example, clothing is designed and used by man for warmth just as fur and hair serve the same purpose in animals. Man is therefore able to reason that animal fur and hair are designed by a Creator for the purpose of providing warmth. Sometimes in this form of the argument man is only partly able to trace God's intentions 'as in various phenomena amenable to the laws of quantity'.

2. Those objects, in which, man sees no more than the preliminaries and the results, or the end and design accomplished; without being able to trace through their details, the means of that accomplishment.

All the phenomena of chemistry fell into this group, while at the other extreme were

3. Those objects, in which, design is inferred, but in which the design, as well as the means by which it is accomplished, are alike concealed; as in the existence of fixed stars, of comets, of organized life.

Chemistry, Prout thought, showed many obvious signs of design, but the actual mechanisms by which these designs were produced could not, unlike the phenomena and objects of the first class, be completely understood by the human mind. Yet the whole burden of Prout's *Chemistry* was to show how macrochemical phenomena could be reduced to the beautiful design of a molecular theory of matter. The total effect, as an *Athenaeum* (1834) reviewer noted, was to reduce chemistry to the design argument of class 1; but this implied that if Prout's molecular theory proved to be incorrect, then his whole argument for the existence of God would be compromised. In fact, however, Prout had cautioned the reader that his molecular speculations were only introduced by way of illustration and that they were not essential to the design argument. In the second edition, to prevent further misunderstanding, Prout stated firmly 'that no argument illustrative of design has been founded on the supposed molecular arrangements which he has given; and that the reality of design in the phenomena of chemistry is no more affected by the truth or falsehood of its imaginary incidents'. Elsewhere, however, Prout suggested that these molecular speculations were correct and 'calculated, sooner or later, to bring chemical action under the dominion of the laws of quantity'.

The theological argument and content of Prout's *Chemistry* followed the pattern of its companion volumes: examples of apparent utility and design in both the inorganic and organic kingdoms were, by cumulative effort, made into an argument for the unity of design and purposeful beneficence of a Superior Chemist. Prout nowhere offered a justification of analogical reasoning. In fact, as Hume had suggested, the use of analogical reasoning in theology has no logical validity for, unlike science, theological hypotheses cannot be tested either directly or by their fulfilment of predictions based on the original hypotheses. Although the use of analogical reasoning in scientific discourse is an invaluable aid to the formulation of hypotheses, these always have to be tested against experience. However, this condition cannot be fulfilled in theology for the design analogy does not produce (in Hume's requirement) a constant conjunction between the hypothetical first cause, God, and his supposed effects, the materials of the world. We cannot confirm the existence of a world designer by watching him make one. Nevertheless, as Hurlbutt (1965: 164) has pointed out (in order to account for the longevity of natural theology after Hume) the design argument is psychologically persuasive because of 'previously acquired and emotionally grounded prejudices in its favor'. In this latter respect Prout's treatise was obviously successful, as a few examples of arguments, both naive and sophisticated, will show.

Although 'the excellent Paley' had remarked that chemistry could not provide the same sort of rigorous argument for a deity that was provided by mechanics, Prout (1834a: 11) maintained that 'the very imperfections and difficulties of chemistry' gave it some advantages over mechanics.

> When the Deity . . . operates through the medium of mechanism, he appears almost too obviously to limit his powers within the trammels of necessity; but when he operates through the medium of chemistry, the laws of which are less obvious, and indeed for the most part unknown to us, his operations have much more the character of those of a free agent, and, in many instances also, appear of a higher order and are more striking and wonderful.

In other words, since chemistry was an empirical science and very much the science of secondary qualities, it was very difficult to discover the causes of chemical phenomena. God was therefore revealed as a free agent in all his omnipotence and not as a God who restricted his power (as in mechanics) through easily recognised laws of nature. For, closely following Paley, Prout (1834a: 16) subscribed to the opinion that the so-called 'laws of nature' were limitations which God had prescribed to his power.

In the actual discussion of the known laws and phenomena of chemistry all theological argument is dismissed, and the treatise becomes an advanced chemistry textbook which deals with the atomic–molecular theory. Eventually, of course, a return is made to the design argument; matter could not have always existed in its present molecular form, and since Prout (1834a: 86) assumed that it could not have attained its designed form by chance, 'it must have been the work of a voluntary and *intelligent* Being'. Prout's programme is clearly similar to Boyle's attempt in the seventeenth century to remove the atheistic stigma from Epicurean atomism.

> And what a still more sublime idea is this calculated to convey to us of the wisdom and power of that Being who contrived and made the whole! When and where do we naturally exclaim, did this Being exist: Whence his wisdom, and whence his power? There is, there can be, but one answer to these enquiries. The Being who contrived and made all these things must have pre-existed from Eternity—must have been omniscient—must have been omnipotent—MUST HAVE BEEN GOD!

This pattern of extravagant rhetoric alternating with long passages of straightforward scientific fact or speculation, is repeated throughout the treatise. Thus, in the fourth chapter, which deals with fifty four of the then known elements, a description of their properties is balanced by a consideration of their functions (i.e. their final causes) as part of the fabric of creation. One example will be sufficient. The French physician Jean Coindet (1820) introduced the use of iodine for the treatment of goitre into French hospitals in 1820; but Prout (1834a: 100) claimed that its use for this purpose had been discovered by him in 1816 'after having previously ascertained by experiments on himself that it was not poisonous in small doses . . . The employment of the compounds of iodine in medicine was at that time made no secret; and so early in 1819, the remedy was adopted in St. Thomas's Hospital by Dr. Elliotson, at the author's suggestion'.

Prout was not embarrassed by the absence of any obvious functions for the majority of the elements; but, taking a tip from Paley, he explained 'who can tell what design is latent under *apparent* incongruities?' The logical implication of a muddled or incompetent designer was thereby avoided. Of course, the fact that many elements were of an injurious or poisonous nature could not be so lightly dismissed. Whereas 'the properties of oxygen stamp it as an element and subordinate agent of the most important kind; while the numberless contrivances which are observable in nature, to secure, or evade, or modify its operations, are most extraordinary, and exhibit some of the most marked an unequivocal evidences of design on the part of the great contriver', the same could not be said of chlorine which was fatal to life in its free state. However, Prout noted triumphantly, chlorine had been really designed to be of use when compounded with other elements—he refers to common salt and the hydrochloric acid of the stomach. This manoeuvre avoided the logically necessary conclusion from analogy that evil poisons had been designed by a malevolent designer.

In this manner Prout developed the casuistical argument of Paley's eighteenth-century optimism that evil was perhaps essential for the greater good of the world. Poisons had been introduced by the creator not because he was unable to prevent their existence, 'but on purpose and designedly to display his power'. Prout (1834a: 169) also proposed a 'scientific' reason, namely that the 'incompatible properties of the elements ... in some way contribute to the perfection of the compounds ...; and the only grounds upon which such incompatibility seems to admit of explanation is, that it results necessarily from those limitations which the Deity has thought proper to prescribe to his power, and to which he always most rigidly adheres'. It had been the properties of compounds rather than the properties of the elements which God had designed 'without reference to the secondary properties of the elements themselves, which were left to be determined as the more general laws of matter might decide'. This argument horrified George Wilson (1862: 27, 46), another later writer on chemical theology.

The Deity induced adaptations in matter by subtle changes of quality and quantity. If the quality (i.e. the composition) of the air or sea had been different, or if it were to alter, then the whole economy of nature would be drastically changed; if the quantity or ratio of land and sea had been different, or if it were to alter, or if the quantities of hydrogen and oxygen present in water were any different, then 'the whole order of Nature would be subverted, and the whole of the present arrangements be involved in ruin'. It is interesting to notice that Prout speculated that such differences might have been responsible for the past upheavals in the earth's history which contemporary directionalist geologists like William Buckland were describing.

The *Meteorology* section contains very little teleological argument. Colour is proposed as an example of God's benevolence, for if snow had been black, absorption of light and heat would have led to severe flooding from melting snow. Fortunately, however white snow absorbs a certain portion of light and heat (by a beautiful provision *more* as the angle of incidence increases?) while so much light is reflected as is useful and no more (Prout 1834a: 242).

The anomalous expansion of water was so important 'that it is doubtful whether the present order of nature could have existed without it', for (in the manner of Cleopatra's nose) if it were not for the comparative lightness of ice, it would have formed from the bottom of water instead of from the surface, and therefore been the last part to melt. Life in lakes and ponds would have been impossible. Prout (1834a: 250, 359) thought that this (unoriginal) argument was one of the most telling examples of design in the whole of nature, especially if it was considered in conjunction with the constant homogeneity of air which was so critical for the existence of terrestrial life.

> In order that ... *water might not be frozen*; and that *air might not become irrespirable; laws must be infringed*—and THEY ARE INFRINGED; infringed too, precisely where their infringement, both in kind and degree, is indispensably necessary to organic existence.

As for climatic effects which appeared to be solely destructive and evil, Prout, like Paley, defended them on the grounds of either the unsearchable ways of God, or of some general benefit.

> A hamlet is laid waste; a few individuals may perish; but the general result is good; the atmosphere is purified; and the pestilence with all its trains of evil disappear.

Even deadly malaria and cholera were evidently designed to stimulate man's reasoning powers to discover antidotes, and to urge him to industry through the conversion of marsh to fertile land. But, generally, Prout seems to have preferred to write about more tractable topics; proofs of benevolent success rather than benevolent failure.

It is, of course, easy to pick out passages from any treatise on nineteenth-century natural theology and hold them up for contemporary ridicule. We must not forget, however, that in a non-evolutionary, providentialist and anthropomorphic conceptual system in which every creature and substance is thought to be designed to play a role subserviant to man, all apparent divergences and exceptions to the fine adaptation of means to an end have to be explained in terms of that system. So to atheists for whom adaptations were merely the result of immanent 'necessary and eternal laws of nature'; and to deists for whom design could not be proved, Prout replied as follows. The so-called laws of nature were founded on either reason and necessity, or on experience. But it was almost impossible to demonstrate any laws of the first kind. The atheist could neither prove his laws were necessary, nor could he prove them eternal. Indeed, for Prout and those who thought providentially, experience suggested that different laws of nature had prevailed in different ages of the world.

> Hence as these laws cannot be proved to have a necessary existence, or to have existed from eternity as they now are, it becomes more than probable that they have had a beginning; and thus the inference of a pre-existent Law-maker, and all its consequences are at once inevitable.

The deist's argument could only be neutralised. As an empiricist, Prout was willing to admit that since we could only understand nature through experience, no *a priori* demonstration of design (and therefore presumably of the existence of God) could be given. But equally, he argued, no *a priori* demonstration that the world had not been designed had yet been given. We should notice that Prout does not say that such a demonstration is impossible, and also that he makes no direct reference to the arguments of Hume who, by the 1830s, was remembered as an historian rather than as a theological critic.

The design argument makes almost no appearance in the section on the *Chemistry of Digestion* until the final recapitulation where Prout maintains that all the facts he had presented were 'obviously demonstrative of design'. The important mutual relations between the inorganic, vegetable and animal kingdoms were a general scheme laid down by Providence. The balanced association of the chemical processes of digestion and respiration with the anatomical structures of digestion and circulation of the blood could only be explained 'on the supposition that *a will exists somewhere*; and also a *power* to *execute that will*' (Prout 1834a: 541).

However, if design is not prominently argued in the third book of Prout's *Bridgewater Treatise*, it is replaced by vitalism, which could be described as microcosmical natural theology. This will be a convenient place, therefore, to look at Prout's vitalism and its relation to his chemical career.

In common with the majority of chemists and physiologists in the first half of the nineteenth century, Prout believed that the chemistry of organised beings was intrinsically different from that of 'unorganised' materials (Brooke 1968; Benton 1974).

> Their form is altogether different, and instead of being bounded by strait lines and angles, it is almost universally made up of some variety or combination of curves [Prout 1816a: 12].

Such differences were to be explained by a 'principle of organisation' or 'living' or 'vital principle', which was present in all organised materials. This governing power, which had been 'endued by the Creator with a faculty little short of Intelligence' (Prout 1816a: 17; Prout 1831a: 261) created, from ordinary inorganic elements and agents, organic substances which were in states of combination different from those found in the mineral kingdom and which 'perhaps also, consist of ultimate and refined forms of matter, which do not naturally, and perhaps cannot exist separately under the present constitution of the universe' (Prout 1816a: 12). There was a continuous battle between the chemical forces and elementary substances of the inorganic kingdom and the organising force of living organisms. The equilibrium of living systems could only be maintained by 'the constant and unremitting agency of the vital principle'. If this vital agent failed either naturally through age, or from sudden exhaustion, then death resulted and 'the *bruta tellus* prevails, and speedily restores the incarcerated atoms to their original state of existence'. This view echoes the vitalism of the French physiologist, Xavier Bichat (1805: 1), who had offered the definition of life as 'the collection of functions which resist death'. However, Prout did not follow Bichat in maintaining that organisation could never be discussed in terms of inorganic forces and substances. While accepting design Prout was also prepared to subscribe to what Lenoir (1982) has called a 'teleological mechanical' framework of explanations.

Apart from the unpublished *De Facultate Sentiendi*, Prout's first physiological essay with a vitalist flavour was the paper on sanguification, published in 1816. In this, as in the *Sentiendi*, Prout maintained that the character of the vital principle could never be understood completely since it was impossible to study it by a comparison of its quality or quantity with any other known object. Here Prout seems to use the phrase 'vital principle' to mean a cause of life and organisation, not that it is some sort of ponderable entity. At other times, however, especially in his later references to the subject, he appears to visualise it as an imponderable or incorporeal agent. Such ambiguity is common in the literature of the period and is merely a reflection of chemists' and physiologists' puzzlement at the phenomena of organised life. On the other hand, Prout firmly believed that the investigation of life by the chemist was not a hopeless proposition, for the animal chemist could investigate the means employed by the vital agency and by the analysis of its general field of activity, map out 'what it can, or what it cannot, or rather what it does, or what it does not, accomplish'. This positive attitude was to be the key to the creation of biochemistry and physiology since it allowed the investigator complete freedom of research.

In his own 'mapping' Prout described two negative features and one positive characteristic of the vital principle. Firstly, the 'principle does not create material elements, or change one element into another' (Prout 1816a: 13; Prout 1834a: 431). In other words, Prout believed that the principle of the conservation of matter, or the conservation of elements, was obeyed in organic chemistry. Nevertheless, we have already noticed in dealing with the chemistry of the egg that Prout was prepared to override this principle and replace it with the possibility that under extreme conditions elements might be transformed one into the other. In fact there is no inconsistency, because to Prout the so-called elements of the chemist were not necessarily the ultimate principles of chemistry; following Humphry Davy, they were merely 'undecompounded bodies'.

> The astonishing discoveries of modern chemistry have reduced the number of elements to comparatively few, and many of those formerly considered as simple substances, are now known to be compounds of two, three, and even more different principles, as for example bone earth, or phosphate of lime. But we cannot suppose, that even yet we are arrived at the limits of our knowledge. Chemistry is still progressive, and as we extend its limits, we diminish the number of our elements, so that even those substances which at this time we consider as simple, may, in the course of a few years, be demonstrated to be compounds [Prout 1816a: 13; Prout 1834a: 431].

On the other hand, it was well known that plants and animals died if they were fed indifferently, whereas if the living principle possessed 'a transmuting power, [the living thing] could subsist on any species of matter foreign to itself'. Since this was not observed, Prout preferred to think that normally the living principle did not exert any creative or synthetic power.

Secondly, the 'vital principle does not combine elements in such a manner, that the result or compound shall differ in its properties from those which it would possess if its elements were combined by any other agent' (Prout 1816a: 13; Prout 1834a: 432). This very important generalisation also made for an empirical approach to animal chemistry. 'Thus phosphate of lime ... the carbonate of lime, the sulphuric acid, the carbonic acid, &c. formed by the processes of organization in animal bodies, differ in no respect from those formed by the chemist'. Chemical laws were invariant whatever the means, inorganic or vital by which they were effected.

> One and one will make two, whether the author of the addition be an ideot [sic] or a philosopher, or even the Deity himself, and the same must be said of a particle of oxygen and a particle of hydrogen, which,

when they unite must make a particle of water, whether the combination takes place by the agency of common chemistry, or by that of the vital principle in organized beings [Prout 1816a: 14].

This meant that the chemist was free to apply his knowledge and his analytical techniques to the field of organised matter.

Thirdly, 'the vital principle operates upon inorganic elements and by means of inorganic agents'. A vital agent directed and coordinated the inorganic forces of heat, light, electricity and 'that power or principle (if it be different from the electric fluid) which is the cause of chemical action' towards the organisation of the inorganic elements carbon, oxygen, hydrogen and nitrogen. Since all these subordinate inorganic agents and bodies were found associated with life, it may be wondered why Prout thought vitalism necessary at all. The fundamental reason was the perceived direction and government of the processes of life.

> It is perfectly ascertained that (heat, light, electricity, &c.), out of an unorganized body, and *left entirely to themselves*, never would or could unite, either in virtue of their own properties, or from accident, so as to form any plant or animal however insignificant. Are we not then compelled to infer, that within a plant or animal, there exists a principle or agent superior to those whose operations we witness in the inorganic? [Prout 1834a: 430].

Whereas today a biologist or biochemist will explain the processes of life by appealing to the action of enzymes or to the influence of a genetic code, the early nineteenth-century scientist, who was without these concepts, explained that a vital agency commanded, directed, and if necessary, modified the actions of subordinate inorganic powers which, in their turn, acted directly upon inert materials according to the known laws of physics and chemistry. Inevitably there were many different positions adopted over the role of vitalism in explanations of living processes. Extreme physiologists like Bichat asserted that because life was such a completely different phenomenon from any inorganic or 'unorganised' phenomenon, the organised could never be explained in terms of inorganic materials and forces. In its extreme form this meant that chemistry was a useless science to the physiologist whose discipline was concerned with 'biological' laws of a fundamentally different order from the laws of physics and chemistry. Such a viewpoint was strenuously opposed by the early practitioners of animal chemistry, Fourcroy, Marcet, Bostock, Prout, Dumas and Liebig, who argued that since life deployed inorganic elements such as carbon in its handiwork, the ordinary rules of chemical affinity were also deployed in living processes. They did not exclude vitalistic explanations, for they conceived that a vital agency (whether material or a force could be left unresolved) commanded, directed and maintained the assembly of inorganic elements within the system. Very few chemists bothered to discuss their use of vitalistic explanations in any detail (Lipman 1964).

Because historians have not always noticed that the majority of chemical vitalists discussed the behaviour of *organised* substances, and not *organic* substances, living in an age when there is a clear, non-vitalist distinction between inorganic and organic chemistry (namely, organic chemistry is the chemistry of the carbon compounds), they have been most troubled and concerned about the question of synthesis and whether the synthesis of a substance like urea, in 1828, destroyed the vitalists' cause (Jacques 1950). This is not to deny that some chemists did assert dogmatically that the synthesis of an organic compound was impossible—as impossible as inorganic synthesis was possible; but such chemists are to be found mainly in the first two decades of the nineteenth century when the problems of purifying and analysing organic substances seemed to make them a class apart from mineral substances and make their obeyance of the laws of definite proportions seem an open question.

Historically, syntheses like that of urea did not destroy chemical vitalism; if this were the case one might well wonder how Prout (1817) could have speculated about the possibility of its synthesis long before Wöhler and yet have remained a vitalist all his working life. The truth is that chemists like Prout were deploying the language of vitalism in order to explain the behaviour of organised living systems rather than of the organic substances which could be extracted from them. The distinction is crucial, for whereas an organised body like a cat, or a tree, or a stomach *in vivo*, is living and vital, an organic body like sugar, or urea, or even albumin, which is a constituent of these bodies, is as lifeless as a mass of zinc oxide. As Thomas Thomson (1820a: 3) wrote:

> Notwithstanding the imperfection of our present knowledge, I see no reason why we should despair of being able hereafter to account for many of those processes which puzzle us at present, and even of being able to form artificially various substances, both animal and vegetable, which we cannot do at present.

Organic substances, made from the same elements found in inorganic chemistry, were the chemical constituents of organised bodies, and it was necessary to suppose the existence of a vital force in the latter in order to account for their formation and stability.

Therefore, Prout was quite in order to speculate and even attempt the synthesis of organic materials such as sugar and urea; but this had no bearing on his vitalism since the synthesis of life itself was by definition out of the question. The vital force might have more efficient ways of producing organic compounds, but there was no reason to think that the chemist would not be able to repeat some of these syntheses in the laboratory. But the synthesis of life itself—of organised materials—seemed quite out of the question.

> The philosophy of chemistry will draw the conclusion that the production in our laboratories of all organic compounds, as long as they are not part of an organism, must be seen as not merely probable but as certain.

So wrote Liebig and Wöhler (1838) in their brilliant experimental paper on the products of nitrogen metabolism. Their prediction began to be verified during the 1840s with Kolbe's work on methods of organic synthesis (Brooke 1971).

Philosophically it may be argued that little was achieved by the postulation of a vital agency; but as Everett Mendelsohn (1965: 203) has pointed out, historians must live with the fact and see how vitalism functioned within the chemical and physiological models used by nineteenth-century scientists. Methodologically, the concept of vitalism proved useful both as an explanatory hypothesis and as a factor which effectively demarcated one science with one set of problems from another with a slightly different set of problems. In this case, the science of biochemistry was encouraged to emerge from physiology and animal and vegetable chemistry.

It should be clear, then, that if Prout had been asked why the vital agent did not act directly upon inorganic elements, or why the subordinate inorganic agencies were necessary, he would have answered that it would be unscientific to have reasoned otherwise. For it would 'give to the vital principle a power little short of a creative one'; it would be like asking a potter to convert clay into porcelain 'without the aid of tools, of the intermediate processes of tempering, baking, &c' (Prout 1816a: 15). Moreover, decay after death was a protracted affair, whereas if the vital agent acted directly on the inorganic elements, then the instant it ceased to operate would see the instantaneous decomposition of the whole organism.

The actual organisation of inorganic materials took place at the minute terminations of the parts, or 'machines', of living things where comminution occurred; later he referred to 'cells' rather than machines (Prout 1845: 389).

> In these minute extremities it is evident, that the principles operated upon, must be separated as it were into atoms, each one of which will thus be capable of being acted upon or not, independently of the rest,

and consequently combined with, or separated from others, in any manner which the economy of the being may require [Prout 1816a: 16].

Prout was here thinking of the formation of the blood in particular, for where the arteries terminated, so also, he believed, did the nervous and absorbent (lymphatic) systems. If, as seemed likely, the vital agent acted through the nervous system by exploiting the inorganic agency of electricity, then the 'synthesis' or organisation of blood and other materials, and the separation of unwanted materials through the lymphatic system, could well take place at this microscopic level. Here, Prout's notion shows some similarity with the later idea of the cellular laboratory.

In the Gulstonian lectures which he gave to the Royal College of Physicians in 1831, Prout began to speak of organic agents rather than a singular agent of separate vegetable and animal living principles. These three lectures on 'The application of chemistry to physiology, pathology and practice' were a continuation of the theme of the 1827 Copley Medal paper, and they also anticipated the discussion in the third book of Prout's *Bridgewater Treatise*.

In his lectures Prout (1831a) anticipated Liebig's propaganda (Holmes 1974) with the argument that physiologists paid too much attention to mechanical and metaphysical explanations in biology and medicine, and none to chemical explanations. Biology and medicine demanded a chemical approach, he stressed. However, it is significant that Liebig (1842a: 378; Liebig 1842b: xiii) also demanded that the chemical approach should be quantitative. Prout (1816a: 289) ascribed the lack of progress in animal chemistry as due both to the intrinsic difficulty of the subject and to the incompetence of the pure inorganic chemist who began to work in the unfamiliar field of biological chemistry. The solution to this problem was for physiologists to become chemists, for they would be well acquainted with the ordinary phenomena of the living state. If they were also trained in chemistry, then they might discover the properties, composition and the conditions of formation and change 'which the mere chemist is apt to overlook, or knows not how to appreciate even if he observes them' (Prout 1831a: 258). This argument was rejected by the physiologist Wilson Philip (1831) in a long, acrimonious correspondence with Prout in the pages of the *Medical Gazette*. It is to be doubted that Prout's argument had much immediate effect in Britain, though Henry Bence Jones (1850), a pupil of Liebig's, later stated that Prout had 'first established the true connexion between chemistry and medical practice'.

Prout continued his Gulstonian lecture with the warning that the chemical physiologist should, after admitting the existence of an organising principle, be a positivist with respect to life itself and avoid all speculation concerning either the nature of life or the vital agency. How, then, were the obvious differences between inorganic and organic substances to be explained? Prout (1831a: 258) here returned to the concept of merorganisation announced in 1827. Using the analogy of the remarkable changes which could be induced in some inorganic substances, such as iron, by the presence of traces of 'impurities', he suggested that the earthy and saline bodies which usually accompany organic molecules were not accidental heterogeneous impurities, but necessary ingredients. Organisation could not proceed without them since they affected the 'energy state' of the complex even though they did not appear to enter into the organic body in any definite proportions. Prout thought the experiments of John Herschel (1824) upon the electrical effects of impurities meant that they acted similarly in organic molecules and perhaps by operating interstitially such particles prevented the organic molecules from assuming the shape of inorganic crystalline forms. In this respect it was significant 'that many of those minute foreign substances which Mr Herschel found to exert most energy in his experiments, are precisely those most usually occurring in organised bodies, such as sulphur, phosphorus, magnesium, calcium, iron, &c'. These materials operated

> by being interposed, as it were, between the essential elementary atoms of organized substances, & thus prevent them from assuming the crystallized form, in which state they would be totally unfit for the purposes of the economy of living organized beings [Prout 1831a: 260].

In the third book of his *Bridgewater Treatise*, these matters are referred to at some length (Prout 1834a: 414; 1845: 396) with little alteration in either words or implication from the introduction to the 1816 paper on sanguification and the Gulstonian lectures. He again argues forcibly that the only way to explain life is to assume the existence of an agency 'different from and superior to, that which operates among inorganic matters'. Organic agents acted in two ways: either by 'peculiarity of composition and of structure', or by virtue of the manner in which this peculiarity of composition and structure was produced. The first method referred to merorganisation, the second to a distinctive way in which this composition and structure was realised. In inorganic chemistry the chemist worked with molecules in the mass, whereas it seemed that the organic agents 'having an apparatus of extreme minuteness, [are] enabled to operate on each individual molecule separately; and thus, according to the object designed, to exclude some molecules, and to bring others into contact'.

From this analysis it seems that the functions of Prout's agents are very similar to those of the 'faculties' of the Greek physician, Galen, or to those of the 'archei' of the seventeenth-century iatrochemist, van Helmont. Like them, Prout effectively required an organising agent for each major apparatus of metabolism, the stomach, the liver, the kidneys, and so on.

> The organic agent, in its simplest state, may be viewed as a power which so controls certain inorganic matters, as to form them into an apparatus, by which it arranges and organizes other matters, and thus effects its ulterior purpose [Prout 1834a: 437].

Where the control of one agent ended, that of another higher, more complex and effective, agent began; and this process continued with ever increasing powers of organisation until 'the perfection of organized existence' was obtained. In the creation of such endless diverse organic agents, God revealed his attribute of infinity.

After reading Prout's Gulstonian lectures, the Manchester surgeon, John Roberton (1831: 745), expressed dismay over the manner in which Prout insisted that organic agents existed in every individual. He could not believe that such an hypothesis was necessary, 'and besides [it] is calculated, by its mysticalness, to retard or discourage the study of this science'. The pragmatic distinction drawn by Roberton between inorganic and organised substances was that the chemist could predict the course of events, or reactions, in the former, but not in the latter where there was a dependence of part and process upon the whole which had no analogy with inorganic chemistry. Roberton was not trying to avoid vitalism (which was a legitimate hypothesis), but he thought Prout's particular brand raised more questions than it answered. What happened to such vital agencies after death? What kind of 'intelligence' did these agents possess? Vitalism, like attraction or affinity, Roberton declared, was only a metaphysical expression to describe a set of facts which did nothing to explain their ultimate causes. For Roberton, organic substances were not intended for manipulation by men, and in opposition to the scientific vitalism of Prout he proposed a theological vitalism in which the sole cause was God.

Prout (who was in the middle of his controversy with Philip) made no direct reply to Roberton, though in the third edition of the urine textbook (Prout 1840: iv), he did add some useful historical remarks on the several hypotheses which had been formulated as explanations of organisation. There was, firstly, the time-honoured transcendental belief in 'independent existing vital principles or agents, superior to, and capable of controlling and directing, the forces operating in inorganic matters'. This was Prout's position. Secondly, there was the suggestion that vitality was not independent of the common properties of matter, but a property superadded to them. This was the position adopted by Liebig and Berthelot and the later organic synthesists; for, as chemistry progressed during the nineteenth century, the need for explanations of the first type was reduced until only explanations of the creation of life retained them. Since this dilute form of vitalism maintained that the 'vital force' itself was open to chemical investigation, Prout had to reject it. To him it was equivalent to saying that material forces like heat possessed the intelligence to select, bring and combine together the primary elements and elaborate them into the 'wonderful mechanism' of an organised being. However:

> material forces fulfil the will of the Creator in organic processes without any knowledge or will of their own; and are mere brutes, which left to themselves, so far from forming organized beings, the moment the vitality ceases, are actively employed in destroying organization [Prout 1845: 401].

Thirdly, there was the suggestion that the simplest form of vitality, irritability, was 'the result of certain aggregations of inorganic matters'. This property, when acted on by other powers, produced more complex phenomena such as sensation and the whole web of actions which were called 'life'. Such an immanentist position had been adopted by the physiologist John Fletcher (1837), who had said

> It is not Life then, but only a necessary condition of Life, namely irritability or vitality which is the result of organism; and when we speak of organized matter, we mean, not that it is endowed with Life—any more than any inorganic matter is endowed with combustion or sensible motion—but only that it possesses a property which, when acted on by appropriate powers, is competent to give rise to that series of actions in which Life consists.

Prout, however, found this as untenable as the former since, to him, it appeared to overlook the element of choice or will found in organisation to preserve existence or to restore health. Laws of matter, or of development were, he thought, incompatable with choice or will; worse still, they compromised belief in the human soul (Jacyna 1983).

Clearly, Prout's vitalism became more theological and metaphysical as he grew older. Whereas as a young man he had been content to describe organic agents as faculties little short of intelligence, at the end of his life he was dogmatically asserting that they were 'conscious beings' possessed with 'knowledge, will, and power'. This sounds dangerously like polytheism—a doctrine which would have shocked Archbishop Howley and Bishop Blomfield, who had presumably vetted the Bridgewater authors for doctrinal deviations. 'We must confess', wrote one hostile critic *a propos* this 'ugly excresence' to an otherwise brilliant book—he was referring to the third edition of the *Chemistry*—'our inability to discern the essential difference between this figment, and the mythical notions of the ancients respecting their deities or semi-deities, which we presume that Dr. Prout would condemn as absurd' (Anon 1846).

Yet, as Prout had pointed out, his hypothesis of transcendental organic agents did not prevent the man of science from investigating organic phenomena. It was an explanation of the simplest kind; one, moreover, which avoided the philosophical, social and political difficulties implied by the assumption of material forces or laws of immanent development.

In this manner Prout fulfilled the conditions of the Bridgewater legacy to the Royal Society. His book combined, in a skilful manner, the argument for the existence of God from design with the sciences of chemistry, meteorology and metabolic physiology. During the course of his exposition, Prout revealed for the first time in detail, as we shall see shortly, his molecular speculations, and he attempted to reduce the phenomena of inorganic, vegetable and organic chemistry to the properties of molecules and agents. In the field of animal chemistry, where he was specifically concerned in the *Bridgewater Treatise* with the problems of digestion, the resulting vitalism was a curious mixture of contemporary organic chemistry with the fashionable 'faculty-type physiology' of contemporary phrenology. In the manner of Galen who, when required to account for a bodily function or process, postulated a 'faculty of alteration', Prout introduced organising faculties, or 'organic agents'. Inevitably there had to be a great number of these agents—as many as there were vegetable or animal functions or processes—and the resulting picture was curiously at odds with his molecular speculations which were always aimed at the reduction or unification of entities. Nevertheless, this kind of vitalism did not interfere with Prout's programme to investigate the chemistry of life; perhaps (as Claude Bernard maintained of his own views later in the century) it had the merit of making it clear that physiology and biochemistry had peculiar problems that differed from those of the physical sciences.

On the other hand, by endowing his agents with intelligence and suggesting that they were immaterial conscious beings, Prout seems to have been unnecessarily extravagant. And his pessimistic opinion that the nature of the vital agency, or life itself, would never be completely understood, has turned out to be a mistaken one. Ironically, it was by exploiting the method which Prout had recommended, the mapping out of what the vital agency did, that permitted the freedom of research in animal chemistry that eventually led to the acceptance of immanentism and the disappearance of vitalism. Otherwise Prout strictly adhered to the laws of physics and chemistry and from hindsight his organic agents can be seen to have many analogies with enzymes and even the genetic instruction book. But although the explanatory functions of organic agents and enzymes, or the genetic code, are similar and sometimes identical, the evolutionary context in which the latter explanations are made has completely replaced the providential and transcendental connotations of vitalism.

Far from preventing or inhibiting Prout from the investigation of the chemistry of digestion, we conclude that his particular metaphysical brand of vitalism allowed him to follow the dynamics of the digestion and assimilation of foodstuffs in a temporal sequence through stomach and intestines, and to make conjectures concerning their chemical and vital transformations into blood or the fabric of the animal body, or the decomposition of the organism and the elimination of waste products. In a word, Prout's vitalism permitted the investigation of metabolism. However, in practice, this programme was only to prove successful when amalgamated with the tools of quantitative and symbolic chemistry. In this step lay Prout's failure and Liebig's success.

5

Prout's Hypotheses

> In a more enlightened period, we have extended our enquiries and multiplied the number of the elements; the last task will be to simplify; and by a closer examination of nature, to learn from what small store of primitive materials, all that we behold and wonder at was created [Chenevix 1803: 320].

The philosophical view that the world of appearances may be reduced to one real invariant (monism) was first examined by the pre-Socratic Ionian thinkers. Empirically, they identified the basic material, or *protyle*, from which the world of appearances was constructed, with a variety of different substances such as water, mist, fire, or more abstractly, with an intangible *apeiron*. Such ideas accepted the phenomena of change and the multiplicity of things without any attempt to explain how they came about. Parmenides produced the logical dilemma of pre-Socratic thought: how could change (including what we would call chemical change) and the variety of material things be explained in terms of one invariant, or element? Historians of Greek thought have recognised that three ways out of this philosophical cul-de-sac were tried. Two of these ways may be described as roads which led to a corpuscular physics, the other a road which led to a non-corpuscular or qualitative physics.

The medical philosopher, Empedocles, argued that change and multiplicity should be explained in terms of several basic roots, or corpuscular, elements which were themselves invariant; i.e. they persisted unchanged through a chemical process, which was therefore really the association or separation of these elements. His qualitatively different elements, four in number, Earth, Air, Fire and Water, were supposed to be omnipresent in both the reactants and products of a change.

The atomists, on the other hand, retained the idea of a single basic stuff (i.e. one element), but they subdivided it into an indefinite number of indivisible and invariant particles called 'atoms' which were only distinguishable by their geometrical primary qualities of shape and size. Atomism may therefore be described as a unitary theory of matter in which a singular element is divided into indivisible parts; these parts, which can only be distinguished by their size, shape and motion, persist through change. Chemical change is reduced to the rearrangement of these atoms in space (the void).

Although atomism was an *a priori* essay put forward to resolve a philosophical problem, whereas Dalton's chemical atomism was proposed to resolve a set of physico-experimental phenomena, it is nevertheless useful to view Dalton's theory, *qua* matter theory, as an amalgamation of these two earlier corpuscular philosophies. Chemical change according to Dalton was explained in terms of a number of qualitatively different (Empedoclean) and quantitatively different (Democritan) atoms. However, these atoms were no longer omnipresent elements, but simply the smallest parts of Lavoisier's undecompounded bodies. By making certain simple assumptions regarding composition—for example, that where only one compound containing two constituents was known, binary composition was occurring—Dalton was able in 1803 to set up a scale of relative atomic masses. Thus, using the symbols and formulae introduced by Berzelius, water was the binary compound HO. If the relative atomic weight of hydrogen was taken as unity, then that of oxygen would be, by the best analyses, 8. This suggestion was adopted enthusiastically by Thomson and Berzelius, both of whom did much to extend and improve the determination of the relative weights of such chemical atoms, the question of their identity with real physical atoms remaining conveniently independent of the usefulness of the calculating system (Rocke 1978).

If we return to the Greeks for a moment, it was Aristotle who first effectively challenged corpuscular explanations of change by denying that it was due to the co-mingling and separation of invariant primordial units of matter. Still faced with the Parmenidean problem, he overcame it effectively with his teleologically orientated doctrines of matter and form, potentiality and actuality. It should be noted, however, that Aristotle retained a first matter which persisted through change and 'carried' the old and new 'forms' through the transformation (i.e. transmutation). This explanation of chemical change, which he never fully developed, was debased over the centuries, especially by the alchemists who treated it as a theoretical basis for their experimentation. The idea became current that the first matter was preparable, and that spectacular and financially rewarding transformations could be performed by exchanges of preparable forms.

Moreover, since Aristotle retained the four Empedoclean elements, the distinctions between the Empedoclean and Aristotelian matter theories became somewhat blurred, and the philosophers in the seventeenth century, who revived a quantitative corpuscular physics, included both those who insisted on a universal substratum of matter (or one element) and those who believed in the existence of several qualitatively different substratums. From the seventeenth century onwards, through the influence of Boyle and Newton, physicists tended to talk in terms of one universal matter, while chemists adhered to multi-element theories. However, as a result of the critical analysis of the theory of matter which Boyle (1661) made in the *Sceptical Chymist*, chemists no longer believed that these elements were omnipresent. Instead, through the influence of Stahl, elements came to be treated pragmatically as those substances which the chemist could not decompose. This received powerful support from Lavoisier in 1789 when he advised chemists against speculations concerning the ultimate constitution of matter, and suggested a purely pragmatic definition of the element.

> All that can be said upon the number and nature of elements is, in my opinion, confined to discussions entirely of a metaphysical nature. The subject only furnishes us with indefinite problems, which may be solved in a thousand different ways, not one of which, in all probability, is consistent with nature, I shall, therefore, only add, that if, by the term *elements*, we mean to express those simple and indivisible atoms of which matter is composed, it is extremely probable we know nothing at all about them; but, if we apply the term *elements* or *principles of bodies*, to express our idea of the last point which analysis is capable of reaching, we must admit, as elements, all the substances into which we are able to reduce bodies by decomposition. Not that we are entitled to affirm, that these substances which we consider as simple, may not themselves be compounded of two, or even of a greater number of more simple principles; but since these principles cannot be separated, or rather since we have not hitherto discovered the means of separating them, they act with regard to us as simple substances, and we ought never to suppose them compounded until experiment and observation have proved them to be so [Lavoisier 1790: xxiii].

By amalgamating corpuscular physics with Lavoisier's pragmatic chemistry, Dalton was in many ways closer in spirit to Empedocles and pre-Boylian chemistry than to the atomism of Democritus. Yet there were many nineteenth-century exponents of a corpuscular chemistry who were closer to Democritus; these adopted the atomism of a universal matter and developed a molecular theory. This school, while it was unable to share the empirical advantages enjoyed by the multi-element viewpoint, nevertheless enjoyed the encouragement of conceptual simplicity and the prospect of a physical, quantitative and mathematical chemistry.

The theory of the elements in the nineteenth century (Knight 1967; 1978) became a dialogue between the school of multi-element chemists and the various schools of reductionists; and since Dalton had built the multi-element viewpoint into his atomic theory, reductionists to some extent inevitably felt obliged to reject simple Daltonian atomism. From the foregoing analysis, however, we can see that in doing this they were not rejecting atomism, or a corpuscular physics; they were rejecting an Empedoclean physics for a Democritan molecular physics.

In the forms known as 'Prout's hypothesis', this viewpoint is attributed to William Prout. However, Prout correctly described it as 'an opinion not altogether new', and for many years historians of chemistry have demonstrated that its immediate intellectual ancestry lay in the writings of Humphry Davy and Thomas Thomson who had both wanted to reduce and simplify the growing number of elements endowed to chemists by Lavoisier and by the labours of Davy himself. The issue is actually a little more complicated, and made more interesting by the fact that Prout's undergraduate essay *De Facultate Sentiendi* shows that he had an Aristotelian-like commitment to a belief in prime matter.

Therefore, before the respective claims of Davy and Thomson to have originated Prout's hypothesis can be reviewed, the evidence revealed by this manuscript source must be discussed. We should first recapitulate the relevant material on matter.

The vital principle, Prout suggested, combined with ordinary inert matter in varying degrees and so produced the characteristics of vegetable, animal and human life. Prout [1810a: 2] took his definition of matter from the eighteenth-century authority on Aristotle, James Harris (1775: 63).

> Matter is that elementary constituent of composite substances which appertains in *common* to them all without distinguishing them from one another.

Consequently, by this Aristotelian definition, all matter was one, or made of the primary matter which had been denominated by early Greek philosophers as $\dot{\upsilon}\pi o\kappa\varepsilon\iota\mu\varepsilon\nu o\nu$. Yet this matter obviously existed in many conditions or forms of secondary matter, the $\ddot{\upsilon}\lambda\eta\ \pi\rho\dot{\omega}\tau\eta$ or 'matter which has a *capacity* for becoming *many* things before it actually becomes any of them'. A third condition of matter was physical, extended and hard. None of these conditions of matter necessarily really existed in nature, for knowledge of them was derived by abstraction and analogy from the secondary or aggregated forms of matter with which we are so familiar through our senses of sight and touch. Prout's purpose in making these distinctions was 'merely to endeavour to render probable by their means the unity of matter'.

The ὕλη πρώτη or secondary matter, was redefined by Prout in a more scientific manner as 'matter in its aggregate state', and by this he referred to such substances as wood, stones, water, air and Lavoisier's chemical elements. Secondary matter could exist in one of five different states of aggregation, solid, liquid, aeriform, atheriform and luciform. It differed from the primary physical form of matter by possessing the quality of roughness.

Inferred	*Perceived*
Prime matter—potential matter—physical matter	Aggregated matter
(extended and hard)	(extended, hard and rough)

As we saw earlier, the essay was largely concerned to argue how the intellectual combination (which was only to be found in man) derived a knowledge of material things through sensation; and it was an attempt through analogical reasoning to reduce all sensational mechanisms to a unity. However, besides this physiological and psychological programme, this was equally an explicit commitment to a chemical programme in which the elements were to be reduced to one kind of matter.

The disordered manuscript notes of Prout's lectures on animal chemistry, which he gave privately in London in 1814, show that he continued his commitment to the unity of matter; likewise he continued to use the Aristotelian–Harris language of 'active' and 'passive' agencies. 'The objects of nature', he wrote in his notes,

> may be divided into *elements* and *agents* or the *passive* and *active* principles. (Corresponding to the pos. and neg.) Ancient and modern conjectures on the nature of the ultimate passive and active principles. General coincidence [of] these in the notion of their being but of *one* kind—i.e. that their [*sic*] is but *one* active principle in nature from which all others are deduced and modified [Prout 1814d: f.1].

If this hypothesis were correct, then the passive primary matter was organised by an active agent into the various elements, compounds, organised beings and higher active agents. Consequently,

> ordinary (?) or compound elements and agents can differ from one another in degree only, i.e. according as the active or passive principles

predominate in their composition and secondly, that the more active principles can act over all the less active ones.

In other words, every natural object was relatively passive (as an element) to certain other objects (agents), but active (as an agent) towards others (elements).

In the same lecture Prout referred to the following diagram.

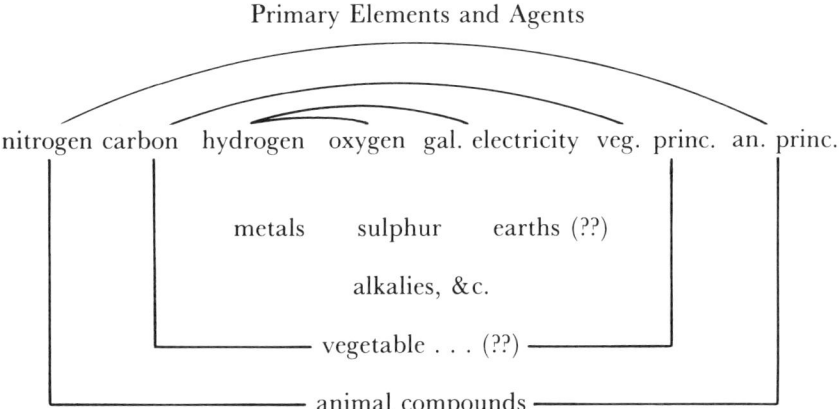

Prout explained that in this table:

> The objects of nature, either as actually known to us, & known by their effects only I divide into two great classes which are denominated *elements* & *agents*. between these two I do not pretend to draw a distinct line, but suppose them to run into each other or in other words that every element is more or less an agent & v.v. those however which from their characters being stronger marked appear better than others to deserve these appellations are placed on the extremities of the scale [Prout 1814d: f.3].

In the division of mineral chemistry, hydrogen was the primary element or passive agent which was converted by the active agency of galvanic electricity or its modification (whatever that might be), into the other chemical elements and agents of mineral chemistry. The other agents were the familiar imponderables, caloric, light and magnetism, and by implication, the vegetable and animal organising principles.

> Metal, of which hydrogen is considered the most characteristic & pure is placed as the primary element—& galvanism with its modification as the ultimate agent—between which two lie all the other elements & agents of this division.

Although oxygen is not mentioned in the text, the diagram (which seems to have been modified for the actual lecture) placed oxygen on the same level as hydrogen, and it is not clear whether Prout supposed that galvanic electricity first acted on hydrogen to produce oxygen, and then that these two elements reacted with the galvanic agency to produce all the other elements; or whether he supposed that oxygen was another primary element which was perhaps as characteristic of the non-metals as hydrogen was of the metals. This ambiguous position over the role of oxygen is also to be found in Prout's first anonymous paper in 1815; it may well reflect Davy's uncertainty over the same question. However, it may be suggested that since Prout speaks in the text of the development of carbon and nitrogen from hydrogen and oxygen, and yet he placed these four elements on the same horizontal line, that he believed that oxygen was also developed from hydrogen. This would be in keeping with his belief that the number of elements was 'doubtless much more numerous than the simplicity of natures operations requires'.

The characteristic primary element of vegetable chemistry was carbon, and the characteristic agent a special vegetable living principle. Since all the resources of inorganic mineral chemistry were also included in this division, the vegetable agency was free to act not only on carbon, but also upon the mineral imponderables, caloric, light, electricity, etc, as well as the ordinary chemical elements. Finally, the characteristic primary element of animal chemistry was nitrogen which was organised by a characteristic animal living principle. Here all the resources of both the mineral and vegetable kingdoms of chemistry were also included.

Prout's lecture notes are unfortunately incomplete, but the general picture is fairly clear. On the grounds of the simplicity of nature, Prout suggested to his audience that an animal principle organised nitrogen and the elements and agents of the other two lower kingdoms into animal substances; the vegetable principle organised carbon and the elements and agents of the mineral kingdom into vegetable substances (and he noted the objection that some vegetable substances had been found to contain nitrogen); and finally, the foundation stone for this simple hierarchy was the mineral kingdom based solely on galvanic electricity and hydrogen.

In one sense this audacious classification could be regarded as a pedagogic device upon which Prout based his lectures. Yet in view of his earlier essay *De Facultate Sentiendi*, and the famous anonymous papers which he published in 1815 and 1816 in Thomson's *Annals of Philosophy*, it is clear that Prout genuinely believed in the complexity of Lavoisier's and Davy's elements. Nevertheless, in his lecture he qualified himself by stating that 'the ocean of hypothesis' was to be carefully avoided. Enough has been said, however, to prove that Prout was committed to the unity of matter while he was a student at Edinburgh, and before the publication of Davy's *Element of Chemical Philosophy* in 1812 publicised the reductionists' cause. The manuscripts from the lecture period of 1814 confirm what has never been seriously doubted by historians of chemistry, namely that Prout had read Davy's textbook since several references are to be found to it. There is therefore no reason to think that this manuscript evidence in any way diminishes the influence which Davy exerted upon the development of 'Prout's hypothesis'. Rather, I believe, it reinforces the impact which Davy had on Prout who was already a convinced reductionist.

The influence of Davy

Robert Siegfried (1956; 1959; 1963; 1964) has pointed out in a series of articles that Lavoisier's pragmatic definition of the element was intellectually disquieting because it left chemists with the problem of deciding whether a chemically undecomposable substance was a real element, or whether time would show that it was compound. Humphry Davy experienced this dilemma in a most acute form, consequently refraining from speaking of 'elements' at all, and preferring the non-commital phrase, 'undecompounded bodies'. Like Richard Chenevix (who was quoted at the head of this chapter), Davy felt instinctively that there were too many 'elements', and he hoped for an explanation of Dalton's laws of definite and multiple proportions in the discovery of bodies which really were undecomposable rather than with an atomic hypothesis that was bound up with the present limitations of chemical analysis in its acceptance of a multiplicity of elements (Brock and Knight 1965). Ironically, of course, Davy suffered the embarrassment of isolating many new Lavoisieran elements, including sodium, potassium and chlorine; yet he remained true to his sceptical reductionist principles and accepted his discoveries as possessing the same ontological status as the other undecompounded bodies. Whether sodium, potassium, the alkaline earth metals, or chlorine and iodine really were elements remained to be seen.

During Davy's 'period of perplexities' between 1807 and 1808, the confusions of his experiments with sulphur, phosphorus and tellurium, in which he found them to 'contain' hydrogen, lent him much support to his reductionist commitment. From 1809 until 1812, Davy made several public pronouncements of his belief in the ultimate simplicity of nature, and these must have awakened considerable interest. In a lecture at the Royal Institution in 1809, he speculated on the probability

> that substances which we at present conceive to consist of different species of matter may ultimately be referred to different proportions of similar species, and in this way the science of the composition of bodies may be materially simplified [Davy 1839–40: viii, 323].

In the same year he wrote to the botanist Thomas Knight, 'I have come to a conclusion ... that water is the basis of all the gases, and that oxygen, hydrogen, nitrogen, ammonia, nitrous acid, &c, are merely electrical forms of water, which probably, according to the αριστον μέν νδωρ is the only matter without power, and capable, according as it receives power, or change in its electricity, of assuming the various forms hitherto considered as elementary' (Davy 1858: 129).

Two years later, in 1811, he publicly discussed the analogous properties of the metals and suggested that they all contained a common elementary principle, or principles. These suggestions reached an even wider public through the appearance of his *Elements of Chemical Philosophy* in 1812. In this Davy (1839–40: viii, 330) wrote:

> A series of proportions may be formed in which the metals may be supposed composed of hydrogen, and another substance in definite quantities; and in this hypothesis, the lightest would contain the largest quantity of hydrogen, and possess as they are found to possess, the strongest attraction for oxygen and chlorine.

In an 'Advertisement' for a projected second edition of his *Elements* which never appeared, Davy (1839–40: iv, xv) drew attention to his adoption of integral proportional weights, and added:

> I have usually given *whole numbers*, taking away or adding fractional parts, that they may be more easily retained in the memory. When the number was gained from experiments in which a loss might be supposed, I have added fractional parts, so as to make a whole number.

Here Davy gave the impression that his sole motivation in adopting whole numbers was to be in keeping with 'an elementary book devoted to the general truths and methods of the science'. Certainly Davy did not try to link this pedagogic simplification in any way with the speculative simplification in the seventh division of the book where he unfolded his methodological commitment to the reduction in the number of elements. Prout, who as we have seen was already committed to a belief in a prime matter, did make such a linkage, and he must have drawn encouragement from Davy's remarks that:

> It is contrary to the usual order of things, that events so harmonious as those of the system of the Earth, should depend on such diversified agents, as are supposed to exist in our artificial arrangements; and there is reason to anticipate a great reduction in the number of the undecompounded bodies.

or that,

> Matter may ultimately be found to be the same in essence, differing only in the arrangements of its particles; or two or three simple substances may produce all the varieties of compound bodies [Davy 1839–40: iv, 42, 132].

and especially Davy's remark that,

> We know nothing of the true elements belonging to Nature; but as far as we can reason from the relations of the properties of matter, hydrogen is the substance which approaches nearest to what the elements may be supposed to be. ... After hydrogen, oxygen partakes most of the elementary character [Davy 1839–40: iv, 368].

This last sentiment was expressed in a similar manner by Prout only two years later.

Davy's mature public opinion was, therefore, that all inflammable bodies contained hydrogen and that oxygen was probably the basis of substances with other properties. However, the exact role of oxygen was left ambiguous. Different electrical states, or different arrangements of the same matter were conceived to constitute different chemical species, just as ice, water and steam were all the same material in different physical states. Even if this hypothesis were shown to be true, the facts and doctrines of chemistry would remain largely unaltered. 'The only change in the science would be, that those substances now considered as primary elements must be considered as secondary; but the numbers representing them would be the same, and they would probably be all found to be produced by the additions of some simple numbers or fractional parts' [Davy 1839–40: iv, 364].

Stated positively, this is of course Prout's hypothesis, and it also foreshadows Prout's later suggestion that fractional multiples of the primary element might be involved. It was along Davy's promising path of enquiry that Prout wandered in 1815.

The influence of Thomas Thomson and John Miers

It was not Humphry Davy, however, whom Prout specifically cited as having first noticed the 'law of multiples', but the protagonist of

Daltonian atomism, Thomas Thomson (1802: i, 386) who, in the very first edition of his popular pre-Daltonian *System of Chemistry* in 1802, had shown a concern for the simplification of the number of elements endowed to chemistry by Lavoisier.

> As the term *simple substance* in chemistry means nothing more than a body whose component parts are unknown, it cannot be doubted that, as the science advances towards perfection, many of those bodies which we consider at present as simple will be decomposed; and most probably a new set of simple bodies will come into view of which we are at present ignorant. These may be decomposed in their turn, and new simple bodies discovered; till at last, when the science reaches the highest point of perfection, those really simple and elementary bodies will come into view of which all substances are ultimately composed. When this happens ... the number of simple substances will probably be much smaller than at present. Indeed it has been the opinion of many distinguished philosophers in all ages, that there is only one kind of matter; and that the differences which we perceive between the bodies depend upon the variety in the figure, size and density of the primary atoms when grouped together.

It is not surprising, therefore, that Thomson became a 'staunch supporter of Prout's hypothesis in later years' (Millard 1955: 18).

There seems to have been some confusion among historians as to what exactly Prout's hypothesis was. In fact there were two hypotheses: one, that atomic weights, or specific gravities were integral multiples of the atomic weight or specific gravity of hydrogen; and two, that hydrogen was the prime matter, or less specifically, that there were only one or two ultimate elements from which all the known chemical elements were constructed. This distinction is important since a chemist could adopt one of these hypotheses without necessarily adopting the other. In order to avoid confusion, I shall always refer to the former hypothesis as 'the integral multiple weights hypothesis', or more simply, the 'multiples hypothesis'; and to the latter as 'the protyle hypothesis' or 'unitary hypothesis'.

Although Thomson clearly supported a reductionist thesis in 1802, he was never to use Prout's multiples hypothesis as a support for reductionism. As we shall see, Thomson's support for 'Prout's hypothesis' was limited to the first empirically testable part of the hypothesis, namely the integral multiple weights hypothesis, and there is no evidence that he ever wrote in favour of Prout's protyle hypothesis.

In the second of an important series of papers *On the Daltonian Theory of Definite Proportions in Chemical Combinations* published in his own journal, Thomson (1813: 114) drew attention to the fact that:

There are eight atoms of simple bodies whose weights are denoted by whole numbers; namely

1. oxygen	1	5. copper	8
2. sulphur	2	6. tungsten	8
3. potassium	5	7. uranium	12
4. arsenic	6	8. mercury	25

An atom of phosphorus is ten times as heavy as an atom of hydrogen. None of the other atoms appear to be multiples of 0.132 [hydrogen]; so that if we pitch upon hydrogen for our unit, the weights of all the atoms will be fractional quantities except that of phosphorus alone.

Thomson drew no generalisation from these integers (apart from the erroneous one that there was no connection between specific gravity and atomic weight) and he appears to have looked upon them only as an additional argument for the oxygen scale ($O = 1$) instead of the Daltonian hydrogen scale ($H = 1$). Nevertheless, it is certain that Prout (1932: 36) read this comment since he specifically referred to Thomson's observation of the commensurateness in atomic weights of some of the metals, while in a letter to Walter Adam, he referred approvingly to Thomson's application of mathematics to chemistry (Prout 1813b).

It is possible that Thomson's use of oxygen as a basis for the scale of atomic weights (which was also supported by Wollaston and Berzelius) may have been a further reason, apart from the influence of Davy, why Prout (1815c) queried 'Is the other factor oxygen?' in his first anonymous paper. Prout could have drawn encouragement from Thomson's work and observations upon atomic weights because he was aware that Thomson shared his and Davy's belief in the simplicity of the world and the complexity of the accepted elements. Once Prout had taken the initiative in 1815 and 1816, Thomson became (as we shall see) his intellectual agent. It is worth emphasising again, however, that Thomson only supported and made propaganda for Prout's integral multiples hypothesis; he held no brief for the protyle hypothesis as formulated by Prout.

There was one other possible influence upon the development of the protyle hypothesis which has not been noticed before. This was the erroneous work of John Miers (1814) on the composition of nitrogen, which he still referred to by its French name, azote.

The analogy between the basic properties of ammonia, soda and potash, and the decomposition of the latter two substances by Davy in 1807, led many chemists to speculate that nitrogen was a compound. At one time, because of the formation of an ammonium amalgam between mercury and ammonia, Davy believed that nitrogen was an oxide of an unknown element, 'nitricum' or 'ammonium'; while, because of the pivotal importance of oxygen in his system, Berzelius did not relinquish his belief that nitrogen was an oxide of an unknown element until 1820. Even Thomson (1814: 139), in his discussion of the nature of nitrogen, wrote 'that azote is a compound body can scarcely be doubted. That it contains oxygen is probable, from its little combustibility'. Nevertheless, he concluded in the manner of Lavoisier, that nitrogen should remain an element 'till some fortunate experimenter succeed in showing us the constituents of azote'.

In a theoretical paper published in Thomson's *Annals* a few months later in 1814, a young jeweller's assistant, John Miers, who later became a distinguished botanist, attempted to reduce the elements to hydrogen and oxygen on the analogy that these two elements were the components of water, nitric acid, nitrogen and ammonia.

> It is needless to point out the several others [i.e. analogies] that must occur to those who investigate the subject, as the whole vegetable and animal world present such numberless instances of wonderful arrangements of the most complex materials formed of a few primary elements by the most simple means that could have been devised [Miers 1814: 370].

If water was a binary compound of hydrogen and oxygen, nitrogen, Miers suggested, might be composed of six atoms of hydrogen and one of oxygen. In modern notation

Water HO
Nitrogen $N = H_6O$
Nitric acid $NO_5 = H_6O_6$
Nitrous acid $NO_3 = H_6O_4$

Nitrous gas (nitric oxide) $NO_2 = H_6O_3$
Nitrous oxide $NO = H_6O_2$
Ammonia $NH_3 = H_9O$
Ammonium $NH_{12} = H_{18}O$.

These speculative compositions agreed with the experimentally determined compositions of these substances; and if it could be shown experimentally that nitrogen were a compound of hydrogen and oxygen, this would lend considerable support to reductionist considerations:

> As it is seen that, by the union of the simple elements with two kinds of compound atoms of a double series, the one formed of a particle of oxygen with one of hydrogen, the other of a particle of oxygen with six of hydrogen, so great a variety of compounds may be generated, a question naturally arises, why may not the atoms of oxygen and hydrogen be capable of uniting in more than these two proportions, and why may not other kinds of matter, at present deemed simple, have also atoms of the same order, but of different members of the same two sorts of elementary atoms? [Miers 1814: 371].

The suggestion that two of the common elements like hydrogen and oxygen, or carbon and oxygen, combined together to form all the known chemical elements, was to become a fairly common variety of reductionist speculation (Carnelley 1886). Miers envisaged molecular units of hydrogen and oxygen which resisted chemical decomposition and which were the known 'elements' such as nitrogen, chlorine, boron, etc. The 'happy state of simplicity' which would result from the verification of this speculation was a challenge which chemists were not to ignore.

> The field is now open for all who feel interested in this enchanting pursuit; the extent of research is boundless beyond conception; and there may probably be gained by the beautiful system of atomic combination a more certain and accurate view into the secret operations of nature than has been obtained by all the valuable discoveries that have enriched the science of chemistry of late years. ... To others far more competent must be left the prosecution of this important task [Miers 1814: 372].

It was Prout and Thomson who took up this challenge.

From the foregoing it will have been seen that there are points of contact between Miers and Prout: both were concerned with the simplicity that would result from a reduction in the number of elements. Miers built his speculations on the prime elements, hydrogen and oxygen, which had been singled out by Davy in his textbook. Similarly, in his first anonymous paper, Prout (1815c) referred to both these elements as the units for arithmetical relationships. Miers adopted, in a clear manner, a molecular theory of matter, and unlike Davy he associated the reductionist hypothesis with proportions by weight. Both these steps were also taken by Prout. Finally, there is a more personal point of contact. Miers lived with his father in the Strand and, until the end of 1814, Prout was his neighbour in Arundel Street. The possibility that Miers attended Prout's lectures in 1814 cannot be proved, but since, like Michael Faraday, he was a young man intent on self-improvement and on an education in the sciences, his attendance at a local series of science lectures is plausible.

There is evidence that Prout had read Miers' papers. In the second half of 1814, Miers (1814: 260) published the details of experiments which purported to prove that nitrogen was a compound of hydrogen and oxygen. He began with water, and on the basis that nitrogen (according to his speculations) contained more hydrogen than oxygen (H_6O), he tried to remove the oxygen from water and transform it into nitrogen by passing steam and hydrogen sulphide through a hot copper tube. Miers obtained a number of very conflicting results including, in one case, the production of ordinary atmospheric air, and in another, the production of a 'sulphureted azotic gas' which caused a black precipitate to form when it was passed into caustic potash. In both these cases Miers claimed that his suspicions of the compound nature of nitrogen had been given a high degree of probability; but as Thomson (1815: 14–15) commented in his annual report on the progress of chemistry in 1814, Miers' results were too inconsistent to prove anything.

Many years later, in 1820, Prout pointed out that the so-called 'pure potash' used by analytical chemists usually contained small quantities of silver, and sometimes lead, iron or copper. This could be shown easily by passing hydrogen sulphide through potash.

> The presence of these metals in the alkalies has doubtless often misled chemists by inducing unnatural appearances. I may mention one striking instance. Some years ago, Mr. Miers announced that he had discovered a gas having the property of precipitating potash and soda black. The gas was a mixture containing sulphuretted hydrogen, and the precipitates arose from metallic impregnation in the alkaline solutions employed [Prout 1820c].

Since Prout stated that his discovery of the impure nature of common potash had been made 'many years ago', it appears likely that he made it as a direct result of reading Miers' papers and repeating the experiments. If this is the case, then Miers' speculations assume some significance, and might go some way towards explaining the diffident manner in which Prout announced the protyle hypothesis in 1815–16.

We have now completed a survey of some of the speculations of Prout, Davy, Thomson and Miers which were proposed before the publication of Prout's first anonymous paper in 1815. We concluded from the evidence of the undergraduate essay *De Facultate Sentiendi* that Prout's belief in the unity of matter was developed from a non-atomic context, the qualitative physics of Aristotle, in which there was a prime matter that gave identity to a world of changing forms. The evidence of the confused manuscripts of his lecture notes on animal chemistry which were delivered in 1814 showed that he retained this commitment to the unity of matter after he had learned of the atomic theory; by which time he would also have been aware of the speculations and suggestions of Davy, Thomson, and possibly also those of John Miers. It is not known when Prout first came to know of Dalton's atomic theory. His chemistry teacher at Edinburgh was Thomas Hope (Partington 1962: iii, 797), an opponent of Dalton's theory, and although Dalton had lectured on his theory at Edinburgh in 1807, it seems unlikely that Prout could have become familiar with Dalton's ideas during his own stay there as a medical student from

1808–11. In none of his manuscript materials for the period 1810 to 1814 is there any mention of the atomic theory. We cannot tell whether Prout read Dalton's textbook during this period (Dalton 1808–10), or whether he reflected on the brief references to Dalton's theory made by Thomson in later editions of his *System of Chemistry*. Even if he had read these sources there is little reason to suppose that he would have grasped the significance of the atomic theory until the publication of Thomson's and Berzelius's appraisals of the theory, together with a review of Gay-Lussac's law of combining volumes, in 1813 (Thomson 1813; Berzelius 1813a). After all, Thomson (1813: 146) made a point of exclaiming in 1813 how little Dalton's theory had been studied in England. From Thomson, and to a lesser extent Berzelius, Prout would have acquired a knowledge of the importance of atomic weights and specific gravities. It was the publications of Thomson and Berzelius, together with the publication of the table of synoptic equivalents by Wollaston (1814) which undoubtedly led to Prout's paper on specific gravities in which the integral multiple weights hypothesis was first formulated. These mathematical relations, which were to attract the attentions of Thomson, were for Prout, as he revealed in 1816, an indication of the unity of matter to which, privately, he had been committed for several years. We must now examine these two famous papers in detail.

Prout's anonymous papers on specific gravity

Prout's first anonymous paper appeared in Thomson's *Annals of Philosophy* for November 1815 with the innocent title, *On the relation between the Specific Gravities of Bodies in the Gaseous State and the Weights of their Atoms* (Prout 1815c; 1932). To single out the unitary hypothesis from this paper (where in any case it is not explicitly mentioned) and its sequel in 1816, is to misjudge Prout's purpose, which was explicitly indicated in the title; namely, a study of the relations between the combining *weights* of substances, and the combining *volumes* of the same substances in the gaseous state. He may well have undertaken the task after reading the recommendation made by Berzelius (1814a: 362), in the conclusion of his long essay on the 'cause of chemical proportions', that research into 'the relations between the specific gravity of a compound body and the contraction which its elements undergo in combining' would be of great value in the development of the atomic theory.

Prout's claim that his observations were 'chiefly founded on the doctrine of volumes as first generalised by M. Gay-Lussac; and which, as far as the author is aware at least, is now universally admitted by chemists', suggests that he was unaware of Dalton's rejection of the law of volumes in 1810 (Dalton 1808–10: ii, Appendix; Prout 1932: 25). If, on the other hand, Prout's paper represented a compromise between the positions adopted by Berzelius and Dalton in their respective treatments of the atomic theory (Russell: 1968) he would hardly have been unaware of this. This fact, together with Prout's treatment of air as a chemical compound, leads one to think that Prout knew very little of Dalton's own opinions at this time. It would probably be more true to say, therefore, that Prout saw himself as an arbitrator, not between Dalton and Berzelius, but between the gravimetric atomic theory as expounded by Thomson and Berzelius on the one hand, and Gay-Lussac's volumetric relations on the other; just as Avogadro (1811) and Berzelius (1813a) had attempted before.

Prout began his paper with a calculation of the specific gravities of the elementary gases oxygen, nitrogen, hydrogen and chlorine. The values of oxygen and nitrogen were obtained from the surprising assumption, contrary to Dalton's views, that atmospheric air was a chemical compound, even though there was no condensation of volume when four volumes of nitrogen were 'combined' with one volume of oxygen in the formation of air. He found a justification for this assumption in the constant composition of air, and claimed that it was an original view. However, the notion that air was a chemical compound had been widely accepted at the beginning of the nineteenth century, and it had been strongly rejected by Dalton (Partington 1962: iii, 765). This seems to be an indication of Prout's lack of training in chemistry, or familiarity with its literature.

Since the volume ratio of nitrogen to oxygen was 4:1, and Wollaston's equivalent weights for these gases were respectively 17.5 and 10, Prout judged that air was a compound of two atoms of nitrogen and one atom of oxygen (i.e. N_2O, or 77.77% N to 22.22% O). Then, by simple algebra, he calculated the specific gravities of the two gases and found them to be in good agreement with experimental values (air = 1).

	Calculated	Experiment
Nitrogen	0.972	0.969 (Biot and Arago)
Oxygen	1.111	1.104 (Thomson

But how did Prout arrive at, or justify, the opinion that air contained two atoms of nitrogen and one atom of oxygen? He must have worked back from the experimentally determined specific gravities which indicated that the atom of nitrogen was specifically lighter than the atom of oxygen. How could this be resolved with Wollaston's atomic weights of nitrogen and oxygen unless either the nitrogen *atom* was split during its combination with oxygen, so that $N_2 = 17.5$, or the volume occupied by an oxygen atom was exactly half the volume occupied by an atom of nitrogen, so that, in modern terms, $O = 8$? The former interpretation, if correct, would have required that air contained two *compound atoms* of nitrogen to one simple atom of oxygen, i.e. in volumes

$$2\,(N_2):1(O)$$
$$\text{or } 4N:1(O)$$

or even $4N:1(O_2)$ where one atom of oxygen occupied half a volume.

However, none of this is explicitly stated, and the only justification for this interpretation is the fact that Prout (1827: 354; 1834a: 132) did 'split' atoms many years later, and at the same time refer to this paper of 1815 as containing the notion of a molecular theory, even though he admitted that it had not been at all clearly stated. It was certainly an understatement for Prout to admit that his ideas had not been clearly presented; in fact the whole effect of his anonymous paper of 1815 is one of intellectual confusion. It is for this reason that I believe that a more favourable interpretation of this passage is that Prout treated the oxygen atom as exceptional, namely, that whereas other gaseous atoms occupied one volume, oxygen occupied only half a volume. This was how Thomson understood this passage. Later, at some unknown date, Prout saw that a simpler, more uniform, system would be produced if atoms (i.e. molecules) were conceived to be at least diatomic. But this would have entailed the adoption of the principle that equal volumes of gases under the same conditions contained the same number of particles—a principle which Dalton rejected, and neither Berzelius nor Thomson were able to accept—because of the electrochemical principle that like-charged particles will repel. Thus, in a significant footnote, Prout (1932: 29) pointed out that since:

> one volume of hydrogen combines with only half a volume of oxygen, but with a whole volume of iodine ... the ratio in volume ... between oxygen and iodine is as $\frac{1}{2}$ to 1, and the ratio in weight is as 1 to 15.5. Now .5555, the density of half a volume of oxygen, multiplied by 15.5, gives 8.61111, and $8.61111 \div .06944 = 124$. Or generally, to find the sp.gr. of any substance in a state of gas, we have only to multiply half the sp.gr. of oxygen by the weight of the atom of the substances with respect to oxygen.

Prout's relationship

$$\text{sp.gr. of gas X} = \frac{\text{sp.gr. of oxygen}}{2} \times \text{atomic weight of X}$$

(air = 1)   (oxygen = 10) is analogous to the relationship

$$\text{vapour density of gas X} = \frac{\text{molecular weight of X}}{\text{molecular weight of H}}$$
$$(H = 1)$$

or

Mol. weight of X = 'twice the vapour density' of X.

Prout's first paper continues, however, to present further problems and signs of confused and hasty composition. It must therefore be seriously doubted whether he realised the full significance of his calculations at that time.

His determination of the specific gravity of hydrogen was of the greatest contemporary importance since, being the lightest known gas, its specific gravity was extremely difficult to determine experimentally. Prout proposed that the traditional method of direct weighing of the hydrogen should be abandoned and replaced by a calculation based upon the specific gravity of a denser compound into which it entered in a known proportion, and whose specific gravity could be accurately determined. Ammonia was ideal for this purpose, and using the information published by Gay-Lussac (1808) that ammonia consisted of three volumes of hydrogen and one volume of nitrogen condensed into two volumes, he calculated algebraically that:

If, 3H + 1N = 2 ammonia,

sp.gr. ammonia = $\frac{3}{2}$ sp.gr. hydrogen + $\frac{1}{2}$ sp.gr. nitrogen

And if x is the sp.gr. of hydrogen, $\dfrac{3x + 0.9722}{2} = 0.5902$

Hence, $x = \dfrac{1.1804 - 0.9722}{3} = 0.0694$

(The specific gravity of ammonia had been determined by Davy as 0.59016, and by Biot and Arago as 0.59; Prout arbitrarily took the average of 0.5902.) Since Prout's calculated value of 0.0694 for the specific gravity of hydrogen was a good deal less than Thomson's experimental value of 0.073, he was naturally jubilant when Berzelius and Dulong (1821) made a new experimental determination of the specific gravity of hydrogen which confirmed his theoretical value (Berzelius 1912–36: i, 188).

It is interesting, incidentally, to notice the effect which this new experimental determination had on Dulong (Berzelius 1912–36: ii, 36)

himself. When asked by Berzelius what he thought of Prout's multiples hypothesis, he commented:

> I will tell you frankly that when Dr. Prout's Memoir appeared, I could not prevent myself from being impressed, even though his arguments were not the sort to lend to conviction. I was the only Editor of the *Annales de Physique et de Chimie*, of which there was then a group, who did not treat this paper as absurdly speculative. I made an Abstract of it for the journal [Dulong 1816]. Since then I have never forgotten this original idea. After you and I had redetermined the densities of several gases I was impressed by the verification of his exactness. As you have seen, the numbers we obtained do approach those which Prout gave. It was above all with the intention of deciding this question that I worked with you on the analysis of air to the precision of a thousandth, in order to know the ratio of the densities of oxygen and nitrogen.

Dulong was referring here to Prout's comparison between the calculated values for the specific gravities of hydrogen, oxygen and nitrogen, and his comment that 'the specific gravity of oxygen as obtained is just 16 times that of hydrogen as now ascertained, and the specific gravity of azote just 14 times' (Prout 1932: 27). If, as Berzelius had suggested, equal volumes of gases under the same physical conditions contained equal numbers of *atoms*, then (on the scale, one volume $H = 1$) equal volumes of oxygen weighed 16, and of nitrogen, 14. In other words, the atoms of oxygen and azote weighed respectively 16 and 14. However, as we have seen, Prout preferred at this time to reason that the oxygen atom only occupied half a volume, so that its atomic weight was only 8.

A similar calculation and adjustment was made for chlorine on the basis of the experimentally determined specific gravity of hydrogen chloride gas. Davy's, and Biot and Arago's values for hydrogen chloride were arbitrarily adjusted from 1.278 to 1.2845 upon the grounds that the lower value was judged 'erroneous in the same proportion that we found the specific gravity of oxygen and azote to be above (which though not rigidly accurate), may yet be fairly done, since the experiments were conducted in a similar manner'. The calculated specific gravity of chlorine was in this way found to be 'very nearly 2.5, in close agreement with Thomson's experimental value of 2.483. In fact, the original value for hydrogen chloride, 1.278 (Davy), would have given Prout a figure, 2.4866, which was in even closer agreement to Thomson's value; however, it would have had the disadvantage that on the scale $H = 1$, it would have given $Cl = 35.83$, which was ambiguously far from an integral number. On the other hand, the adjusted figure for hydrogen chloride used by Prout (1932: 28) gave a value of 36.02 for chlorine, or exactly 36 times that of hydrogen'. So much for calculations prearranged to conform to an hypothesis! It is quite obvious that Prout deliberately chose values for the specific gravities of substances which would produce integral numbers on the hydrogen scale; and he can only have done this because he felt that these numbers had some sort of physical significance. Since Prout's paper was so liberally sprinkled with such dubious qualifying phrases as 'just', 'does not differ much from', 'exactly', and even the oxymoron, 'correct, or nearly so', it is surprising that hardly any of the bitter criticism which was later thrown at Thomson was applied to Prout. As Kendall (1949–52) commented, 'No element, it is obvious, could fail to fit into a scheme of such flexibility'.

In a second part of his paper, Prout (1815c) extended his treatment to the solid elements, iodine, carbon, sulphur, phosphorus, calcium, sodium, iron, zinc, potassium and barium. This extension is remarkably similar to the extrapolation which Avogadro (1814) had made to the applicability of his hypothesis. Most of Prout's estimations involved quantitative analyses of his own devising, and they were not without originality and skill, despite Thomson's feeling that they were open to various inaccuracies (Thomson 1816a: 19). However, the procedure still remained flexible, and convenient specific gravity values were always chosen. For example, Gay-Lussac's value for the atomic weight of iodine was 156.21 ($O = 10$), but Prout considered this too high and lowered it to 155. This gave a specific gravity of 8.61025 which Prout printed as 8.611111 (*sic*), or 124.06 times the specific gravity of hydrogen; in Prout's terms, 'exactly 124'.

It is obvious, therefore, that although according to the title of his paper Prout promised to calculate specific gravities from atomic weights, in fact he chose atomic weights which would produce integral specific gravities when compared with hydrogen. The net effect was the correction of atomic weights by means of an integral weights hypothesis. In the case of his own experiments which 'were made with the greatest possible attention to accuracy, and most of them were many times repeated with almost precisely the same results', Prout (1932: 31) was clearly justified in making such adjustments since he knew the experimental error involved. Equally clearly, he was not justified in adjusting the experimental results of other chemists.

All the results were summarised in a table which showed the calculated gaseous specific gravities ($H = 1$), atomic weights (one volume $H = 1$), atomic weights ($O = 10$) from calculation, the calculated specific gravities (air $= 1$), the experimental specific gravities (air $= 1$), the weights in grains of 100 cubic inches of the substances at 30 inches pressure and 60°F, and finally, the experimentally determined grain weights of 100 cubic inches. Two other tables exhibited 'the proportions, both in volume and weight, in which they [the elements] unite with oxygen and hydrogen'. However, by extending his calculations to compound gases Prout introduced (like Dumas later) a confusion of nomenclature since he omitted to qualify the word *atom* by the admittedly paradoxical description *compound*. Worse still, by not clearly stating the principles of his molecular philosophy, Prout (1932: 36) muddled together *atoms* and *volumes*.

> [The] table also exhibits one or two striking examples of the errors that have arisen from not clearly understanding the relation between the doctrine of volumes and of atoms. Thus ammonia has been stated to be composed of one atom of azote and three of hydrogen, whereas it is evidently composed of one atom of azote and only 1.5 of hydrogen, which are condensed into two volumes, equal therefore to one atom.

Prout's argument appears to be that since according to Gay-Lussac

1 vol nitrogen + 3 vols hydrogen = 2 vols ammonia

that the two volumes of ammonia had to be replaced by *one* compound atom of ammonia; hence the three volumes of hydrogen had been distributed or condensed into one atom of ammonia. Prout's *weight* equation was, therefore

1 atom nitrogen + 1.5 atoms hydrogen = 1 compound atom ammonia.

Nevertheless, in the corrected table published by Prout (1816b), the weight equation was given as

1 atom nitrogen + 3 atoms hydrogen = 1 compound atom ammonia

and no reason for this particular correction was given.

In a final table Prout grouped together the specific gravities of the common metals as calculated from Berzelius's atomic weights. Significantly, all of them were integral like the common gases and non-metals of the first table. Prout (1932: 37) commented that:

> all the elementary numbers, hydrogen being considered as 1, are divisible by 4, except carbon, azote, and barytium [i.e. barium], and these are divisible by 2, appearing therefore to indicate that they are *modified* by a higher number than that of unity or hydrogen. Is the other number 16, or oxygen? And are all substances compounded of these two elements?

The significance, if any, of Prout's divisors 2 and 4 is obscure, and one wonders what the contemporary reader made of the expression 'modified by'. If the divisors represented submolecular units, this would be equivalent to saying that molecules comprised two or four particles. I believe this to be the correct interpretation, but since none of Prout's contemporaries could have known this, they must have found this passage incomprehensible. Even if by the phrase 'elementary numbers', Prout referred to both the atomic weights and the specific gravities on the hydrogen scale, there seems to be nothing in the tables to have justified Prout's query whether oxygen was the other fundamental substance. This makes it highly probable that he was referring implicitly, or obliquely, to Davy's, or possibly Miers', speculations. It should be noted, however, that there was no explicit formulation of the protyle hypothesis in this paper; this followed in the second paper.

Instead, Prout picked out for principal comment the integral values of the atomic weights on the hydrogen scale; he noted with satisfaction that: 'Substances in general of the same weight appear to combine readily, and somewhat resemble one another in their nature'. Ernst von Meyer (1891: 190) in his *History of Chemistry* accused Prout of arbitrarily altering the numerical values of atomic weights 'so that they should not merely be whole numbers, but should also show regular differences among each other'. This seems to be true, since a number of atomic weight values were in arithmetical series.

Ca	Na	Fe	Zn	Cl	K
20	24	28	32	36	40

No doubt privately Prout thought of 'missing' elements, just as Mendeleeff was to do much later. This also underlines the fact that Prout, like Thomson, was searching for a periodic law as well as a theory of the elements.

Prout closed his paper with a final enigmatic generalisation that 'all the gases, after having been dried as much as possible still contain water, the quantity of which, supposing the present views are correct, may be ascertained with the greatest accuracy'. This conclusion did not follow from the tables; it is little wonder that Dulong (1816: 416), and later Kopp (1873: 379), found this remark extremely puzzling. Verbosity can be a great failing in scientific literature, but equally, extreme terseness can lead to obscurity. It could be argued that Prout was playing the role of Thales: all things are water (i.e. hydrogen and oxygen), and how much water they are can be estimated. But a more likely explanation is that Prout saw that the amount of water vapour in gases could be estimated by some sort of correction factor once the ideal specific gravities had been calculated.

It is a relief to turn away from these obscurities and failures of presentation to Prout's short correction of 1816 which, though still not entirely free from ambiguity, is a model of clarity in comparison with the first paper. The muddle over atoms and volumes was resolved when Prout proposed that the advantage of relating or correlating atoms with volumes, by the assumption that one volume of hydrogen was occupied by one atom of hydrogen, was that,

> the specific gravities of most, or perhaps all, elementary substances (hydrogen being 1) will either exactly coincide with, or be some multiple of, the weights of their atoms [Prout 1932: 40].

If, on the other hand, the assumption was made that one volume of oxygen was occupied by one atom of oxygen (i.e. 2 vols H = 1 atom), the relation would not be so simple, for then 'the weights of the atoms of most elementary substances, except oxygen, will be double that of their specific gravities with respect to hydrogen'. Unknown to Prout, this assumption had been proposed already by Avogadro (1811); it was to be made by Prout again in 1834 and is equivalent to the classical chemical relation:

$$\text{molecular weight} = 2 \text{ (vapour density)}.$$

This relationship appears to have been the source of Prout's error in the table headings published in 1815.

However, in the 1816 correction Prout remained orthodox and employed the first, simpler relation to demonstrate how the specific gravities of elements in the gaseous state could be rapidly calculated by means of the Wollaston slide-rule. The famous protyle hypothesis followed immediately:

> If the view we have ventured to advance be correct, we may almost consider the πρώτη ὕλη of the ancients to be realised in hydrogen; an opinion, by the by, not altogether new. If we actually consider this to be the case, and further consider the specific gravities of bodies in their gaseous state to represent the number of volumes condensed into one; or, in other words, the numbers of the absolute weight of a single volume of the first matter (πρώτη ὕλη) which they contain, which is extremely probable, multiples in weight must always indicate multiples in volume, and vice versa; and the specific gravities, or absolute weights of all bodies in a gaseous state, must be multiples of the specific gravity or absolute weight of the first matter (πρώτη ὕλη) because all bodies in a gaseous state which unite with one another unite with reference to their volume.

That such an opinion was not altogether new was, as we have seen, a reference to Davy, and perhaps to his own private thoughts. It should be noticed that Prout no longer referred to the multiple of oxygen; unless the guarded qualification that hydrogen was *almost* the protyle of the ancients was an oblique reference to other primordial units of matter. However, this is equally likely to be a guarded reference to the suggestion which Prout allowed to be published in 1831 that the ultimate unit of matter might be a submultiple of hydrogen (Daubeny 1831: 129). Until then, Prout's contemporaries understood him to mean that the absolute value of the specific gravity represented the number of volumes of hydrogen which had condensed together to form a particular element.

$$n \text{ volumes } H_x \longrightarrow (H_x)^n$$

where, if $n = 6$, the element was carbon, if $n = 14$, the element was nitrogen, etc, and x is an integer which until 1831 was taken to be 1.

Although Thomas Thomson (1816b: 343) was the first to reveal Prout's identity in print in the May issue of his journal in 1816, he had said earlier that he had a good idea of the anonymous author's identity (Thomson 1816a: 17). This would not have been difficult for him since as editor of the *Annals* he would have been familiar with Prout's erratic handwriting from Prout's previous submissions for publication. Prout (1816a: 50) himself drew attention to his authorship in the June issue of the *Annals of Medicine and Surgery* in the same year, and not in 1817 as is usually stated (Partington 1962: 713). This suggests that he personally gave Thomson permission to reveal his identity.

It is intriguing to notice how, after his initial diffidence about the authorship, Prout (1827: 354; 1834a: 133) became quite proud of these papers since he frequently (and sometimes unnecessarily) referred to them in the course of his later publications. The reasons for his anonymity remain obscure; the 1815 paper was neither his first publication, nor his first venture into chemistry, and he was already well known in London as a physician and animal chemist. It was Stas (Prout 1932: 42) who said: 'Prout had so little faith in the exactness of his hypothesis that he published it under the veil of anonymity'. But in his lectures in 1814 Prout had evidently not hesitated to reveal his commitment to the belief in the unity of matter, and in any case this was a fairly common assumption. It would appear, therefore, that the anonymity was connected with his contribution to the atomic theory, rather than the speculation concerning the complexity of the elements. Perhaps he felt unqualified to make a contribution to the atomic theory because it appeared to be the research territory of the theoreticians and experimentalists, Thomas Thomson and Jöns Berzelius. He certainly sufficiently distrusted his own proficiency as an experimentalist (where he actually excelled) that he took the bulk of his experimental data from 'superior(s) to himself in chemical experiments and fame' (Prout 1932: 25). The protyle hypothesis cannot be a sufficient reason for the anonymity since it did not figure explicitly in the first paper, and if it had not been for the errors in this paper, Prout presumably would not have published the hypothesis at all at this date.

As we have just seen, Prout announced publicly in 1816 that since he had demonstrated, largely by calculation, that the atomic weights on the hydrogen scale were whole numbers, possibly all the so-called elements were really polymers of hydrogen. The simplicity for which many natural philosophers had been searching since Lavoisier had defined the concept of the element was over; the embarrassing status of Lavoisier's elements was resolved; back to Thales, everything was hydrogen!

Such an hypothesis was obviously exciting, and from our twentieth-century hindsight we realise that there was more than a hint of truth in the idea. Potentially, the hypothesis could be tested, and therefore, after a slow beginning, catalysed by Thomson, it inevitably stimulated analytical work (Rainy 1826; Johnston 1832). For if chemists could show by exact and improved methods of analysis that atomic weights were integral multiples of the atomic weight of hydrogen within the limits of experimental error, then this would provide a means for correcting atomic weight values and support (but not prove) the view that everything was hydrogen. Proof that elements with integral atomic weights were made of hydrogen could only be given if these elements were decomposed with the evolution of hydrogen. Once more we notice the importance of clearly distinguishing between the integral multiple weights hypothesis and the protyle hypothesis; the latter was a world view or conceptual system that was really quite independent of the empirical verification or falsification of the former. One other fact also becomes clear. If Prout's protyle hypothesis were found to be true, then the status of Dalton's multi-element atomic theory in which there were as many different atoms as elements would be threatened. An atomic theory of matter would still be possible, but the simple theory which Dalton had proposed would have to give way to the older corpuscular physics in the form of a molecular theory of matter. Inevitably what happened in the nineteenth century was that a dialogue took place between the school of multi-element chemists and the school of reductionists; and one result of this dialogue was the adoption of a molecular theory during the 1850s and 1860s.

What was the effect, we must now ask, of new improved chemical experimentation upon this dialogue? A very general answer may be given first. Analysts tended to support Prout in Great Britain and America where the principal spokesman was Thomas Thomson, but reject it on the Continent where the principal spokesman was Berzelius. (Like all generalisations, this one provides a notable exception—Jean Baptiste Dumas.) Essentially no public pronouncement on the controversy was made by Prout until 1827, and to all intents and purposes he appeared to have retired into the field of biochemistry. That he had not entirely lost interest in the matter will appear in a later chapter.

6

Prout's Molecular Theory and Biochemistry

> Chemistry forms the connecting link between that kind of knowledge which is founded on quantity; and those kinds of knowledge which rest solely on experience [Prout 1845: 490].

Content to allow others to test the validity of the multiple weights and protyle hypotheses, after 1816 Prout devoted all his research attention to biochemistry. He recalled later that his views on the relation between specific gravity and chemical proportions:

> led me to others which I was exceedingly anxious to verify; and as I was interested ... in the composition of organic substances, it struck me that by submitting these substances to analysis, I might not only obtain a knowledge of their composition, but by investigating the laws which regulate the union of the elements, hydrogen, carbon, oxygen and azote, be able to obtain an insight into the laws which regulate the union of other elementary principles [Prout 1822a: 424]

The result was a highly speculative matter theory which 'after twenty years of close attention and no ordinary labour, we have been induced to consider as the most simple and consistent with the phenomena' (Prout 1834a: 22). Prout published two slightly different versions of this molecular theory in the first and third editions of his *Bridgewater Treatise*, 1834 and 1845 respectively. The later version was given in much fuller detail and it is undoubtedly a development of the 1834 thesis as well as an elaboration of it. In addition a draft of the long 1845 version, which dates from 1837 and offers some additional details, has survived in manuscript. A composite version of this theory will now be described, and this will be followed by an attempt to link the theory with the account of Prout's chemical and physiological work which has been presented in earlier chapters.

Although a large number of hypotheses had been proposed concerning the nature of molecular forces and their interaction, including the much discussed mathematical ether theory of the Italian astronomer Ottaviano Mossotti (1836), which Prout had read about in the *Connexion of the Physical Sciences* by Mrs Somerville (1842), he found them all too complex and 'utterly unworthy to be ascribed to the Deity, whose primary laws are all simple, general, and comprehensive'.

Since the time of Newton, Prout explained, natural philosophers had recognised the operation of two antagonistic forces in their analysis of reciprocal motions: an attractive centripetal or gravitating force, and a repulsive, centrifugal or opposing force. Both forces belonged inherently to every atom and aggregation of matter within the universe, the force of repulsion being usually recognised as the 'inertia' of matter, i.e. its resistance to the attractive force. Prout (1845: 37) noted that the alternative 'dynamic physics' of Kant and the Naturphilosophen (whom he did not identify) 'while it does not exclude Newton's principles; may, from its more general character, be said to comprehend or include them'. In this mechanical system, the repulsive force resulted from the motion of matter which was conceived to possess an inherent tendency to revolve on its axis with a velocity inversely proportional to the mass. The attractive force, on the other hand, was directly proportional to the mass, and exceeded the repulsive force by an amount that was recognised as the attractive force of gravitation. It was this system which Prout adopted in his own brand of molecular physics where he also pursued another aspect of the Newtonian etherial tradition, the unification of forces. In Prout's case this comprised the reduction of heat, light, electricity and magnetism (the classical imponderable fluids) to attractive and repulsive forces produced by the axial rotation of minute masses of matter. His most likely mentors here were Davy, Faraday and Oersted (Cantor and Hodge 1981).

According to Prout's polarity theory, matter was composed of spherically-shaped molecules which possessed a natural 'tendency to revolve on their axes' with a velocity inversely proportional to their molecular weight. God had created two antagonistic forces between molecules: an attractive centripetal force proportional to the mass, and a repulsive force inversely proportional to mass which resulted directly from the resultant motion. Prout agreed with Mossotti (1836) that the force of gravitation was probably the excess difference between these forces, for if it were not for 'the opposing centrifugal force, the whole of the matter in the Universe would instantaneously rush together in one mass'.

Like Berzelius, Prout explained cohesion and chemical affinity in terms of polarisation (Russell 1963). Polarity, caused by the 'motions of two contiguous molecules on their axes', was of two kinds: *homogeneous* polarity, or cohesive attraction or repulsion between similar molecules; and *heterogeneous* polarity, or chemical attraction or repulsion between dissimilar molecules. In the two 1834 editions of the *Chemistry* Prout only described his ideas concerning homogeneous molecular arrangements, i.e. what we would describe today as the physical or molecular forces between the molecules of a homogeneous material.

In the solid crystalline state, Prout conjectured, identical spheres which each possessed a polar, or chemical, axis that was electrical in nature, and various magnetic cohesive axes, united together in three-dimensional agglomerations. Davy's statement that 'the laws of definite proportions and the electrical polarities of bodies, seem to be intimately related' (Davy 1839–40: iv, 40), was combined by Prout with the recent work on electromagnetism. The fundamental experiment of Oersted (1820), which had established the existence of magnetic forces distributed in a circle in the neighbourhood of an electric current, was treated as a model of the situation at molecular level. The chemical axis of a molecule was analogous to the wires carrying the electric current. But in the case of the wires, as Ampère (1820) had shown, in consequence of the magnetic forces which were also present, similar wires were attracted or repelled according to the direction of the current. These effects were analogous to the action of cohesive axes in the molecule. Among molecules, attraction occurred if the equatorial motions were in opposite directions, for their identical velocities would be dampened and a static union produced; repulsion occurred if the equatorial motions were in the same direction. It seemed 'very probable, nay almost inevitable ... that the electric polarities correspond with the supposed chemical polarities, and the magnetic with the cohesive polarities of our molecule'.

Prout clarified his model by referring to the following diagrams. The identical spheres,

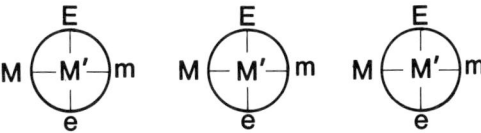

in which Ee is the electrical (chemical) axis, Mm, M'm', etc are equatorial magnetic (cohesive) axes, can stabilise their mutual rotatory forces by aggregating into a cubic mass, or some other derivative solid.

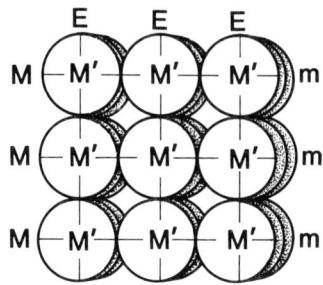

This was clearly analogous to Ampère's electromagnetic experiment in which parallel wires were attracted when the electric currents were flowing in the same direction.

Like all etherial theories (Cantor and Hodge 1981), Prout's polarity hypothesis was a model which easily accommodated changes of phase. In the first published version of the hypothesis, he treated heat as the polarisable ethereal fluid, caloric. As in Dalton's atomic theory, he supposed that caloric formed an atmosphere around a particle, and inasmuch as the overall volume of the molecule was increased, the forces of magnetic cohesion were weakened so that a rotation of the molecules about their axes of polarity easily occurred. This 'loosening of the bonds' explained both the production of, and the different physical properties of, liquid and gaseous states of matter. However, in the more detailed later version of his hypothesis, Prout (1845: 501) described 'the molecules of the imponderable fluids light and heat ... as *immeasurably smaller*, and to move with inconceivably *greater velocity* than the molecules of any ponderable substance'. This idea seems to have followed a speculation by Davy (1839–40: iv, 67) in 1812 that

> in etherial substances the particles move round their own axes, and separate from each other, penetrating in right lines through space. (i.e. self-repulsive) Temperature may be conceived to depend upon the velocities of the vibrations; increase of capacity of the motion being performed in greater space; and the diminution of temperature during the conversion of solids into fluids or gases, may be explained on the idea of the loss of vibratory motion, in consequence of the revolution of particles round their axes, at the moment when the body becomes fluid or aeriform, or from the loss of rapidity of vibration, in consequence of the motion of the particles through greater space.

Prout's theory, then, was a compromise between Davy's kinetic theory and the old imponderable theory of caloric. At the same time it extended Davy's idea of molecular rotations and vibrations to produce electric and magnetic polarities. Thus, in Prout's fully developed molecular theory, heat was no longer an etherial atmosphere which surrounded an atom but a beam of swiftly rotating particles which directly interfered with the normal velocities, and hence polarities and orientation of ponderable molecules.

However, the chemist was principally concerned with the interaction between *different* molecules. In heterogeneous polarity, unlike contiguous molecules were in rotation. Since such molecules differed in weight and volume, they differed also in angular velocity; their interactions were, therefore, slightly different from those of homogeneous contiguous molecules. If two different contiguous molecules were similarly orientated, they could not become mutually stationary and cohere because their equatorial rotations were disparate. Instead static cohesion could only take place if their axes were in a vertical plane as

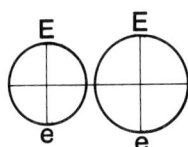 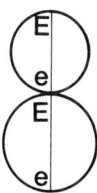

Yet this cohesive state could never be continuous since 'two dissimilar molecules may unite statically [only] at those moments when the motions of the two molecules are coincident; e.g. two molecules, the one moving twice as fast as the other, may unite at every revolution of the slower molecule, which will be coincident with every second revolution of the quicker'. This analysis enabled Prout to suggest that the most stable chemical combinations should be formed between substances which were most simply related by weight. This hides the assumption that atomic or molecular weights are simply related, i.e. Prout's multiple weights hypothesis. For if the atomic weights of the elements are whole numbers these numbers will be (presumably) proportional to the angular velocities of their molecules. For example, hydrogen and oxygen form the stable combinations H_2O and H_2O_2 because the velocity ratios $1:8$ and $1:16$ are reasonably simple. None of this analysis was made explicit by Prout, and it would be unfair perhaps to suggest that velocity ratios like $1:8$ and $1:16$ are not of the simplest kind; or that this explanation of stable combinations is a case of *post hoc ergo propter hoc*. Prout's molecular model, at least in principle, has the makings of a good explanatory and predictive theory provided that measurements of molecular angular velocity can be made independently of atomic weights.

Finally, Prout explained that if two heterogeneous molecules had their chemical axes reversed there would be no possibility of any kind of union since this would be prevented at the equators by similar rotational directions, and at the poles by opposing velocities or polarities

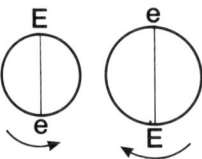

One possibility remained, however, namely that the dissimilar molecules might be so arranged that their axes were at an angle. In this case an equilibrium might be possible, especially if the axes were at right angles. Such a combination 'is very unstable, and favourable for change'.

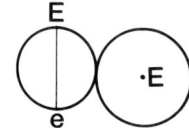

Prout did not mention the conditions which might produce such a union, but presumably the agency of heat would have been required to distort the axes.

It is clear that in Prout's theory homogeneous and heterogenous union are very similar mechanisms. In fact reduction was possible, he concluded.

> Heterogeneous or Chemical Union is only a particular case of Homogeneous Statical Union; and that Heterogeneous or Chemical Repulsion is the same, or rather results from the same cause as Homogeneous Divulsion.

From this he drew the important conclusion that single molecules could not exist by themselves, but that they could only exist in at least binary groups. This had significant consequences for Prout's interpretation of organic chemistry.

But Prout (1845: 161–3) wanted to go still further towards the unification of forces. Light, electricity, gravitation and chemical affinity, he argued, were all manifestations of an attractive force; while heat, magnetism and cohesion all arose from a repulsive force. The differences between these individual phenomena were only apparent and arose simply from 'the differences in magnitude and consequent different intensities of action among molecules'. This is rather vague, but once again, although this was never explicitly stated by him, it seems clear that a full mathematical development of his theory was dependent upon accurate values for atomic and molecular weights.

As for the experimental macrophenomena of sensible electricity, electrolysis and magnetism, Prout thought that these arose from the operations of various grades of molecules: electricity from the very smallest molecules, electrolysis from molecules of a higher order, and so on. The immediate cause of such phenomena was the separation of contiguous molecules so that they moved 'together in the same direction'. When groups of molecules moved together in opposite directions, positive and negative polarities were produced whose intensities were of two kinds.

> the intensity depending on the greater or less velocity of the molecules; and the intensity depending on the greater or less separation of the poles of contiguous molecules; ... the *quantity* of polarity is proportional to the *number* of pairs of molecules moving together at the same time and in the same direction.

In the manuscript version of his theory Prout was a little more explicit. Since molecules were at least binary, when a group of molecules moved in one direction an equal number were forced to move in an opposite direction. Such movements could only take place in the liquid and gaseous states for the symmetrical molecules of solids were rigid and immovable.

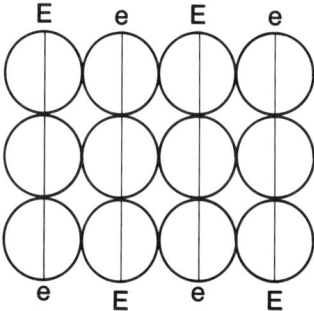

Normal homogeneous gas

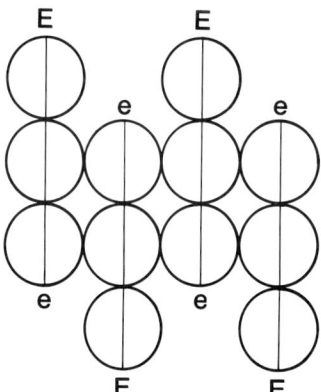

Polarised homogeneous gas

Sensible static electricity could arise from the joint motions of any kind of molecule, ponderable or imponderable, but since the intensity of polarisation depended on the size and velocity of molecules, the sensible electricity produced by the motions of the imponderable fluids heat and light were much more intense than anything produced by the motions of, say, oxygen or water. Heat and light, or rather the minute swiftly rotating molecules whose interaction with other molecules was perceived as heat and light, were therefore probably the principal constituents of static electricity and magnetism respectively. Galvanic electricity, on the other hand, seemed to 'depend on the intensely excited action (by ordinary electricity) of the molecules of oxygen and hydrogen, &c'. Thus, just as the interaction of the smallest particles with larger molecules produced heat, light, magnetism or electricity, these new states might in turn, through further interaction with the ponderable elements, produce the phenomenon of electrolysis and induction.

Although William Whewell (1847: ii, 355) dismissed Prout's model as an arbitrary speculation, it had one important consequence for Prout's molecular theory of chemistry. In his treatment of the change of phase, solid to liquid to gas, Prout placed the cohesive axes into positions of extreme repulsion. In this manner he was able to derive the familiar Daltonian self-repulsive gas particle; a model which agreed with the experimentally known facts of the homogeneity of the atmosphere, the phenomenon of gaseous diffusion, and the well established gas laws. But Prout (1834a: 62; 1845: 99) went further than Dalton by noticing that the gas laws suggested that 'all gaseous bodies under the same pressure and temperature contain equal numbers of self-repulsive molecules'. This is Avogadro's hypothesis.

Although Prout did not deduce the equal numbers of molecules hypothesis from his polarity model, this model was nevertheless an effective explanation and justification for it, since it explained the experimentally known gas laws. For:

> every molecules of matter, when it is in the gaseous state, and subjected to similar pressure and temperature, may, without reference to its other properties, be supposed to be in circumstances exactly similar and consequently liable to be affected in an exactly similar manner by all further increments of heat.

It is intriguing to notice that Avogadro (1811) had not deduced his hypothesis in this way from the identical behaviour of all gases under changes of temperature and pressure, but purely from a consideration of Gay-Lussac's law of combining volumes (Morselli 1984). There seems to be no particular reason to doubt Prout's statement that he arrived at the molecular law 'long before he was aware of the essays on the subject by Messrs. Avogadro, Ampere [sic], and Dumas' (Prout 1834a:64). He does not appear to have studied the English translation of Ampère's molecular theory which appeared in the *Philosophical Magazine* in 1815, and he stated that he learned of Ampère's suggestion only through a review of the state of chemistry by Johnston (1832). But since Johnston's review did not mention Ampère, but Dumas's *Traité de Chimie appliqué aux Arts*, it appears that Prout must have studied this book (where Ampère is mentioned) after reading Johnston. Prout also cited the rejection of the hypothesis by Michael Donovan (1832) as the source of his knowledge concerning Avogadro. Gaudin's molecular theory was not noticed by him at all (Gaudin 1833).

Later, in the section of his *Chemistry* devoted to the laws of chemical combination, Prout (1834a: 122) argued that the equal numbers of molecules hypothesis also followed logically from Gay-Lussac's law of combining volumes which 'established that bodies, in their gaseous state, combine both chemically and cohesively with reference to their volumes'. The polarity hypothesis implied that homogeneous cohesion could only take place between a binary system of molecules at the very least; and so like Avogadro, Prout noted that in the formation of steam from hydrogen and oxygen, 'every self-repulsive molecule has been divided into two, and consequently must have originally consisted of at least two elementary molecules, somehow or other associated so as to have formed only one self-repulsive molecule'. It is ironic that Prout, like Avogadro, avoided the term atom, and used the same terminology as him, namely 'elementary molecule' instead of atom. Other common gases, like chlorine and nitrogen, were also binary molecules 'united to each other cohesively and acting as a single one'.

The conclusion that the molecules of elementary bodies in the gaseous state were binary was, of course, very important and significant. For on this assumption Prout overcame the traditional electrochemical objection to Avogadro's hypothesis that similar parts would repel one another, and therefore never cohere. However, the model which justified this assumption was too much for a Daltonian to follow, as Prout's opponent William Charles Henry (1834) was quick to point out. For it seemed to a Daltonian atomist that caloric brought about a change of state not on the individual atom (as in Dalton's original theory) but on a bi-molecule, or arbitrary group of atoms. As an orthodox Daltonian, Henry believed that each atom attracted caloric to itself and in this way exerted a self-repulsive force. Quoting Laplace in support, Henry argued that he could not see how such interatomic forces could be resolved with the formation of stable, yet self-repulsive, groups of atoms. According to orthodox theory, Prout's molecules would simply disintegrate!

> The contrary hypothesis of Dr. Prout involves the anomaly of supposing heat to have a combining affinity for *two or more* atoms, while it is *destitute* of such affinity for *single atoms*; and also that of supposing two atoms to have relations towards two atoms, or three towards three, which do not obtain between single atoms.

This was a powerful argument against Prout, and it could only be avoided by the adoption of a different heat model, such as the one used by Prout in the later version of his theory. Indeed, Avogadro's hypothesis was to be first generally adopted by physicists, who derived it from a mechanical, kinetic theory of heat.

In reply Prout (1834b) repudiated the implication that he had been beaten to the post by other chemists, by claiming that his 1815 anonymous paper had been founded on this molecular hypothesis, 'understood but not expressed', and that 'from that time to the present I have seen no reason to doubt its truth'. Only a mathematical demonstration that it was incorrect would convince Prout that he was wrong, and he reiterated his Gulstonian lecture statement (Prout 1831a: 262) that there were several details of Daltonian atomism that he found impossible to accept. Indeed, he had always thought Dalton's work on atoms much less satisfactory and complete than his work on gases.

> The atomic theory of Dalton by connecting chemistry with quantity was undoubtedly the greatest step that has been made in modern times; but ... My notion of the atomic theory is, and always has been, that it does not represent a just view of the laws which regulate the union of natural bodies, and consequently that it is inapplicable both to organic and inorganic chemistry. The light in which I have been always accustomed to consider it has been very analogous to that in which I believe most botanists now consider the Linnaean system; namely, as a conventional artifice, exceedingly convenient for many purposes, but which does not represent nature. On the continent, the modification of Dalton's views, proposed by Berzelius, is generally adopted, but this I fear, is still more imperfect than our own.

Daubeny (1831: 62, 131) had thought when he read this in 1831 that it read like a censure motion against Dalton, but Prout had hastened to inform him that he had a great respect for Dalton. He subscribed to Dalton's postulates, as far as they went, he said, but he was not satisfied that they went far enough. In other words it was Dalton's naive atomism that Prout was against, and he would have replaced it by his own more complex molecular theory which was based on volume relations, and in particular on Avogadro's law. In this respect, Prout claimed, Berzelius's volume theory was as bad as

Dalton's for it too did not go far enough. Nevertheless, at an elementary level or first approximation, Dalton's system would probably continue to be used. The danger here was, however, that atomists would continue to fall 'to the temptation of adapting the results of their experiments to the standard set forth by the theory of definite proportions'; that is, by restricting the types of combination to prevent the division of the 'atom' in a chemical reaction. This had been the failing of the three giants of atomism, Dalton, Thomson and Berzelius.

Unfortunately, Prout declined to reply to Henry's critique at any length or depth. His reputation in England and on the Continent as a chemist and physician was considerable, and a technical defence of his ideas in an important journal like the *Philosophical Magazine* would, without a doubt, have been very influential.

To all intents and purposes, then, Prout reached the same conclusions about the relations between combining weights and volumes as Avogadro before him; however, like Avogadro and Ampère, Prout did not build into his molecular hypothesis any rule of simplicity. We have seen already that Prout described gaseous molecules as composed of 'at least two' parts.

> The self-repulsive molecules of oxygen and hydrogen are at least double; but the probability is that they are in reality much more compounded, as the following observations will show. The self-repulsive molecule of water, on entering into combination, is often found to be divided into two, or three (perhaps more) parts. Now as we cannot admit the division of the ultimate molecule, or atom; we must of course conclude, that the molecules of oxygen and of hydrogen, are much more compounded ... and must each of them contain at least three components, or *submolecules*. Hence the self-repulsive molecules (viz. three of oxygen, and six of hydrogen) which we may suppose to be associated, in the first place, the hydrogen with the oxygen *chemically*; and afterwards the three submolecules of water with one another *cohesively*, so as to constitute one spherical molecule [Prout 1834a: 125; 1845: 134].

The simplest volume equation therefore appears to be

$$2H_6 + O_6 \longrightarrow 2H_6O_3 \longrightarrow 2\,[3(H_2O)]$$

rather than

$$2H_2 + O_2 \longrightarrow 2H_2O \longrightarrow 2H_2O.$$

At the back of Prout's mind here, and following logically from a unitary theory of matter, was the series theory of proportional weights which he had briefly mentioned in the Gulstonian lectures in 1831.

There, in his polemic against Dalton, he had argued that the numbers conventionally labelled atomic or equivalent weights:

> appear to me often nothing more than one term of a natural series peculiar to each body, and determining its combination. Thus 9, the number assumed to represent the combining weight of water, is to be considered only as one term of the series 3; 6; 9; 12; 15; &c., in all which proportions (and perhaps in still lower submultiples of them) this fluid enters into combination, perhaps as often in the proportion 9, especially in the organic kingdom. Chemists have already a glimpse of this important fact when they speak of bodies uniting to others in the proportion of two, three or more atoms, which, in fact, are nothing more than the terms of a natural series such as that alluded to [Prout 1831a: 263].

In other words, since the equivalent weight of water is 9, it only enters into reactions or combinations, in sub- and super-multiples of three equivalents. Stability is given to the molecule by the homogeneous forces. By implication elements also have a natural series which determines their combining powers. The water series, 3; 6; 9; 12, was followed in combinations with carbon (e.g. in the saccharines), but other series were possible when water combined with substances which had a different combining series.

> Thus in a natural group or family as the saccharine group ... by adhering to a single number, as 9, for water, we should be led to fractions of atoms without end, but by considering the carbon as associated with different proportions of water, in terms of the above series (as experiment indicates to be the case) all these absurdities are avoided, and at the same time the existence of a beautiful law is indicated.

Daubeny (1831: 44) was inclined to think that this was a needless complication of the law of definite proportions; but since Prout's presupposition of homogeneous matter was different from Dalton's position, this was not the whole story.

In the *Chemistry* Prout insisted that his molecular hypothesis was the simplest and most consistent explanation of a wide range of phenomena. One example, that of isomorphism, will suffice here.

The single cohesive force, which Prout (1834a: 131) had adopted to explain crystallisation, could also be used to account for isomorphism; for 'when the molecules of different bodies are of the same size (or rather of the same weight), they may be naturally supposed capable of associating themselves into the same form; and if they happen to be mixed together, they may even enter indiscriminately into the same crystal'. This is a correct explanation for isomorphism. In his Gulstonian lectures Prout had recorded the opinion that the principles of Dalton and Berzelius were totally incapable of explaining the phenomena of isomorphism and isomerism; and in his letter to Daubeny of the same year Prout (1831b: 131) claimed to have hit upon their true explanation in terms of weight and size relationships in 1815

> So long ago as 1815, I was led to infer that relation in *weight* might indicate a relation also in *size* among the atoms of bodies; and that many of those striking and curious analogies in property, form, &c. which I thought I observed among bodies atomically related, might depend upon one or other of these circumstances.

I think that Prout (1932: 36) was being completely honest here, for he did make brief remarks in the 1815 paper on the relationship between atomic weight and chemical properties. 'Substance in general of the same weight appear to combine readily, and somewhat resemble one another in their nature'. According to the polarity hypothesis, different elements were of different sizes and polar intensities; weight therefore was correlated with size. This suggests that Prout may have derived the polarity theory and molecular law of Avogadro during the period 1815 to 1818.

Prout's explanation of isomorphism led him to speculate that when molecules in a solution did not possess the necessary shapes and angular velocities for satisfactory cohesion, they probably made up the required shape, or acquired the right velocity, by attaching themselves to other kinds of molecules. This 'completion' was the role of water of crystallisation which, for example, enabled molecules to readily combine together and form symmetrical crystals. But other molecules could also act as completing agents, and Prout thought that this was an explanation of many types of isomerism which, it will be recalled, he called merorganisation. In his biochemistry Prout developed this idea to describe the completion of organic molecules by minute quantities of interstitial 'impurities'. In particular, Prout (1834a: 425) suggested, much of the great variety of substances in organic chemistry was due to the presence of minute quantities of strongly self-repulsive foreign molecules equally dispersed among the more ordinary molecules which strongly modified their molecular arrangements and their observed chemical and physiological behaviour.

The molecular theory and biochemistry

In previous chapters we have seen how, between the years 1815 and 1823, Prout devoted much of his attention to the biochemistry of digestion. The publication of a book on this subject was interrupted by, among other things, the discovery in the summer of 1823 of hydrochloric acid in the gastric juice of animals. From Prout's remarks it seems evident that this discovery opened up an entirely new world of research for him, and made him intensify his attempts to apply the molecular theory which he had developed to the physiological problems of digestion and metabolism. In this pioneering field Prout attempted unsuccessfully, without the aid of chemical formulae, what Liebig and his pupils began to accomplish with success during the last decade of Prout's life.

It seemed to Prout (1827) that his analyses had revealed the existence of arithmetical relations between the various members of the three alimentary bodies which he had described as the saccharinous, albuminous and oleaginous bodies. These relationships were especially clear among members of the saccharine family (Prout 1817; 1834a: 483) or vegetable aliments, which were all composed from carbon and the elements of water, and which all formed oxalic acid or its analogues with nitric acid. Some of these materials like sugar, vinegar and lactic acid existed in crystalline forms; others like starch, lignin and the gums were uncrystallisable and, in Prout's language, more organised. He firmly rejected Liebig's belief that the saccharine aliments were used solely to produce animal heat (Prout 1845: 441; 1848: 459), an opinion with which Liebig was later forced to agree. Prout believed that saccharinous substances were transformed into both fats and albuminous substances, but this raised the problem of the source of the extra nitrogen. Albuminous substances or 'animal aliments' were never found in crystalline forms except when drastically modified, e.g. uric acid. All albuminous substances contained nitrogen but Prout (1845: 374) never went beyond this to venture an opinion concerning their chemical constitution except to dismiss Mulder's protein hypothesis for its crudity. He did not deny that a common molecule could be recovered from albumin, fibrin, etc, but 'that a substance obtained like proteine by the rude and disorganising processes of common chemistry, should be that common proximate element; or that such a substance should ever be employed at all in vital operations without undergoing the preliminary assimilating process, is more than at present we are disposed to admit'. All parts of an animal's body contained albumin and gelatine which could be easily separated by boiling with water. Gelatine was the least organised kind of albumin and in many ways the analogue of the saccharine principles in plants for it could be broken down into sugar just as starch could.

Although Prout (1848: 461) stated that he had made many analyses of oleaginous substances no such analyses were ever published by him. He believed, however, that they were binary compounds of ethane (olefiant gas) and water. Oils were usually divisible into two portions, a stearine and an oleine.

Most organised substances contained at least three primary elements; but, Prout insisted from his polarity theory, they must be dualistically arranged. Thus in the sugar series, the binary molecular aggregates were carbon and water, and in the oleaginous family, ethane and water. The spherically arranged forces of homogeneity were extremely strong in organised substances, and this was an addition effect due to their intensity in the primary elements carbon, oxygen and nitrogen. These intensities explained the peculiar ability of carbon to form very large supermolecules of 'atomic' weight 12, 18, 24, etc.

What then was the biochemical significance of the arithmetical relationships revealed by the analysis of members of the alimentary families? Cane sugar was composed of nine atoms of carbon and eight atoms of water, each associated by cohesive forces into two supermolecules weighting $(9 \times 6) = 54$ and $(8 \times 9) = 72$ respectively.

	C	Water
Lignin	54	54
Cane sugar, wheat starch	54	72
Sugar of honey, arrowroot	54	108
Absolute vinegar	24	27
Crystalline vinegar	24	36

The above analytical table, which was published by Prout in 1834 (Prout 1834a: 483), suggested to him that the carbon proportion (or supermolecule) remained constant in saccharines while the proportions of water varied. Vinegar belonged to a different carbon series, and therefore exhibited different properties than the sugars.

Since water was the only variable within a given carbon saccharine series, it could be called appropriately the 'modifying supermolecule' of the sugars. But the chemist was neither able to convert cane sugar into honey sugar by adding the requisite amount of water, nor able to perform the reverse procedure without decomposing the honey sugar. Therefore, Prout concluded, water must be present in a state of 'essential union' in these sugars.

He thought that similar relations could be established in the fatty or oleaginous family, where the basic unit was olefiant gas (CH_2), and in gelatinous or albuminous substances. But it was a great weakness of Prout's biochemistry that he should have been so vague concerning the detailed chemistry of these families; in fact he never produced any examples of their arithmetical relationships.

After he had established the existence of the series relationships among saccharinous substances Prout (1834a: 487) introduced the terminology of the sugar refiner: all *strong* or *high* organic substances had constituent supermolecules of a less complex kind than those of *weak* or *low* substances.

> in the *strong*, fixed and solid oils or fats, ... the modifying molecule of water is very small, perhaps in some oleaginous bodies, is even a submolecule. Whereas, in alcohol, which is the *weakest* condition of the oily principle, the weight of the modifying supermolecule of water is more than half that of the olefiant gas, and alcohol is perfectly soluble in water.

But weak substances possessed less intense homogeneous forces than were found in strong substances in which these forces had become saturated.

The conversion of a strong compound into a weak compound was to be called 'reduction' (i.e. the addition of water); and the reverse process whereby water was removed, 'completion'. Two generalisations followed immediately, namely that the greater the molecular weight, the more easily was a substance decomposed, for the homogeneity was low; and the greater the amount of water contained (i.e. the greater the weakness), the greater was the solubility.

Weak substances  Strong substances
(complex, high   (less complex, low
molecular weight,   molecular weights,
quite soluble,   quite stable)
potentially unstable)

On first acquaintance it might be thought that Prout misapplied the two terms completion and reduction, for they are usually taken to imply, on the one hand, bringing or raising to entirety or fulfilment and, on the other hand, lowering to the simplest. However, Prout's usage is immediately clarified by its application to the biochemistry of digestion.

In the stomach substances were reduced to their weakest forms by the combined action of hydrochloric acid and other water secretions. This was a purely chemical process, but as was remarked in Chapter 3, Prout increasingly came to look upon hydrochloric acid as a pathological disturbance and by-product of chlorine, the real 'powerful influence' that brought about reduction. The chlorine or hydrochloric acid was derived from sodium chloride in the blood which also produced the alkalinity of the bile secretions. The mechanism by which this separation was effected was by the electrolysis of blood.

We have in the principal digestive organs, a kind of galvanic apparatus, of which the mucous membrane of the stomach, and perhaps that of the intestinal canal generally, may be considered as the acid or positive pole; while the hepatic system may, on the same view, be considered as the alkaline or negative pole [Prout 1834a: 496].

Reduction was the principal chemical feature of digestion, or 'primary assimilation', as Prout preferred to call it, in contrast to 'secondary assimilation' which included both the processes of tissue formation from the blood (formative process) and the destruction and removal of unwanted parts from the animal system (destructive process). Secondary assimilation was, therefore, equivalent to what Liebig called the 'metamorphosis of tissues', and in Prout's view necessitated the existence of disorganising agents. The blood itself was formed from chyle and lymph produced during primary and secondary processes of assimilation respectively.

To some extent reduction was performed artificially by human beings during the cooking process, but 'unfortunately, cooks are seldom chemists ... hence their labour is most frequently employed, not in rendering wholesome articles of food more digestible ... but in making unwholesome things palatable' (Prout 1834a: 493). Since the reducing function of the stomach was easily deranged it was of the greatest importance for the dietician to understand this function. For if the function was weak it would be foolish to give a patient heavy, meaty solids instead of pulpy foods; if the function was intense, as in diabetes, then solid, hard, but nutritious, foods were called for. These nutritional points, as well as Prout's notion that milk was the great alimentary prototype on which all diets and cookery should be modelled, were repeated by Liebig, who somehow gained the credit for them.

Two other functions of the stomach were to convert (we should say synthesise) one class of aliments into another when necessary in order to produce a chyle of uniform content (conversion), and the organisation and vitalisation of the food materials. Whereas conversion was a purely chemical process, Prout held steadfast to the belief that the other process was vital. But his belief in chemical conversion meant that, several years before Liebig and Dumas, he had suggested that sugars could be converted into fats, and even into albumins; and fats into albumins, and vice versa. The question of the possible conversion of carbohydrates into fats became a very controversial issue during the 1830s; it was supported by Liebig's German school, but rejected by the French school of animal chemists (Holmes 1974). The matter was eventually settled in favour of Prout and Liebig during the early 1840s. Not until 1843 was sugar detected in the blood, and Bernard's discovery of glycogen was not made until 1856.

As regards the ability of the animal body to convert one kind of foodstuff into another Prout's molecular and unitarian theory of matter led him to a sensible assessment. However, it also led him to expect that under certain unspecified conditions the synthesis of absent elements (like nitrogen) might take place (Prout 1848: 475). This was an assumption that Liebig, and to a lesser extent Dumas, rejected. Usually, Prout thought, the nitrogen came from a nitrogenous source that was already present in the blood and which was secreted into the duodenum during primary assimilation. The non-nitrogenous part of this secretion was then extracted from the blood by the liver or stomach as lactic acid. Prout offered no chemical evidence whatsoever in support of this conjectural mechanism.

In the duodenum the chyme from the stomach was mixed with bile and pancreatic juices and its acidity neutralised by the re-formation of sodium and potassium chlorides. Although Prout (1819a: 273) had observed that the bile induced a precipitation of the fluid contents of the duodenum, Prout, along with Dumas and Liebig, saw no reason to believe that fats were in any way chemically altered in the small intestine. Not until 1849 was it shown by Bernard that pancreatic juice was indispensible for fat absorption (Holmes 1974). Prout thought that the excremental portion of the food was separated out in the duodenum and the remainder of the chyle absorbed by the lacteal system.

Once the chyle entered the lacteals, the opposite process of completion began whereby the strong, and now transformed and vitalised, aliments were passed into the general bloodstream, while the excess water was passed to the lungs for release during the respiratory process. Prout offered no evidence that water was transpired from the lungs in this way; in fact this suggestion, which had been made originally by Lavoisier and Fourcroy, had been disproved by Allen and Pepys (1808). But Prout did not follow Lavoisier's other suggestion that the carbon dioxide exhaled from the lungs was a product of the oxidation of carbon in the lungs. Instead, he ascribed the carbon dioxide to the reduction of albumin in the tissues to form the gelatine that was found in connective tissues like the skin. In order to explain the presence of albuminous substances in the 'absorbents', namely the tissues served by the blood stream, Prout (1834a: 534) had to postulate that gelatine was reconverted back into albumin. He failed to mention that carbon would be necessary for this reverse process to occur; and Frederick Holmes (1965) has argued that Prout failed to understand that since the same amount of carbon would be needed for albumification as for gelatification, there would be none left to be exhaled. However, Prout (1845: 473) does seem to have been aware of this defect since later he qualified that only certain (unspecified) forms of albuminous substances could form gelatine, and that there was only one source of carbon dioxide production. He was also uncertain whether carbon dioxide production was correlated with the production of animal heat and, unlike Liebig, he did not argue that heat was the central function of animal metabolism.

Clearly the greater part of Prout's metabolic system was more ingenious than experimentally clear and detailed. In any case it must be remembered that he did not believe that the processes of reduction and completion were simply chemical; vitalising, or organic, agents were supposed to be distributed in both the stomach and lacteals.

The general vagueness, or qualitativeness, of Prout's account of primary assimilation is also found in his treatment of secondary assimilation, even though this was given in more detail in his later clinical treatise *On Stomach and Urinary Diseases* in 1840. According to Prout, as has already been stated, secondary assimilation was both formative and destructive. These two processes were conducted by the hydrolysis of active molecules into either a new and important metabolic principle or (more commonly) into two principles, one of which was excremental. When two products were formed the molecules were complementary. Sometimes, especially in derangements, three excrementious products might be formed. For example:

$$\text{albumin} + \text{water} \xrightarrow{\text{normal process}} \text{gelatine} + \text{hydrated carbon dioxide}$$

$$\text{gelatine} + \text{water} \xrightarrow{\text{abnormal process}} \text{ammonium carbonate}$$

$$\text{gelatine} + \text{water} \xrightarrow{\text{normal process}} \text{saccharine} + \text{urea}.$$

Gelatification or gelatinification was a formative process by which the albuminous substances within the blood stream became the solid gelatinous tissues that formed the foundation for all other animal tissues (Prout 1848: 484). It took place in the capillary blood vessels and was the direct cause of the transformation of arterial into venous blood. Albumification or albuminification was the other formative process whereby water was eliminated and albuminous substances within the blood were converted into solid albuminous tissues. The process included the fibrification of fluid fibrin into the solid muscular fibrine. In contradiction to Liebig, who held that nitrogenous foods were not heat-procuring, Prout supposed that both these processes were sources of animal heat, but in neither case did he present any experimental evidence to support his account of these fundamental changes. Nor did he have anything to say by way of explanation of formative processes like the synthesis of bone and hair, or the production of internal secretions like saliva and semen.

In destructive secondary assimilation, tissues were unmade and converted either into new useful materials, or into simple and usually crystallisable excretion products. These processes were very delicate and easily went wrong, leading to pathological conditions. For example, gelatinous tissues were normally resolved into the complementary principles of urea and lactic acid which were then excreted through the kidneys or the skin. Sometimes, however, other abnormal decomposition products were formed, or the normal products were produced in excessive amounts; this was then indicative of disease. The products of the destruction of albuminous tissues were rather problematic. Uric acid was certainly one product, though it was most usually to be found excreted in the form of ammonium urate. No evidence was given by Prout for the production of urea and uric acid from the degradation of gelatinous and albuminous tissues. On this theory, one would expect muscular exercise to destroy albuminous tissues faster than gelatinous, and that urine would contain more uric acid than urea. But among mammals the opposite is found.

Prout (1848: 499) summarised the theory of nutrition which was to be the basis of his pathology as follows.

> Water, the basis for all organisation, is first combined with carbon to form the primary organised compound the *saccharine radical*; that the saccharine radical by undergoing certain changes is converted on the one hand into the *oleaginous radical*; and on the other, by undergoing certain changes and involving azote, is converted into the *albuminous radical or principle*. Consequently, *that the albuminous principle on account of its involving the inferior radicals, must in all its ulterior changes be more or less influenced by the presence of these radicals.*

Since milk, the alimentary prototype, contained all four natural aliments, water, saccharinous, albuminous and oleaginous, as well as certain important mineral matters, Prout recognised that the chemical basis of pathology could be approached by considering the normal assimilation and excretion of each class of aliment. This approach was the burden of his revised textbook on stomach and urinary diseases.

Water was used in human metabolism both as a solvent or diluent for certain chemical processes, and more intimately in the association with organised principles in reduction. In the former type of function water was directly absorbed into the blood from the stomach and gut, and used to maintain the fluidity of the blood, or as a solvent for various noxious products that were excreted as perspiration from the skin and as urine from the kidneys. Associated water was usually taken into the body with the food, or as combined hydrogen and oxygen. The ability of the stomach to absorb water directly was easily deranged, and the presence of excess water in the stomach led to imperfect digestion of the food, much discomfort and distress. Such sufferers were advised by Prout to avoid pulpy foods or drinking with their meals.

It was possible to differentiate the urine produced by primary and secondary assimilation by examining urine before and after a meal. Urine collected immediately before breakfast was representative of secondary destructive assimilation (*urinas sanguinis*), and its abnormalities shed valuable light on pathological states such as nephrites and diabetes.

Prout devoted most of his attention to the pathology of saccharine assimilation and excretion. Unfortunately, his ignorance of glycogen and the exact function of the liver invalidated most of his discussion within a few years of his death. Since he believed that normal healthy blood did not contain sugar, he supposed that the saccharine principles had to be completely transformed during their primary assimilation into oleaginous and albuminous substances. If the stomach or other digestive organs were unable to accomplish this very important function, diabetes resulted. However, since in diabetes reduction was, if anything, enhanced, it was only the converting and organising functions that were abnormal. A similar explanation was given for the production of other abnormal substances like oxalic acid which were usually symptoms of a future calculus. Sugar might sometimes also be produced during secondary assimilation when gelatine was produced from albumin, and vice versa. This only happened during the most advanced states of diabetes. More commonly, lactic acid was produced which immediately predisposed a patient to urinary complaints. The best treatment was a diet containing the minimum of saccharinous or farinaceous materials; but one which was carefully blended with vegetable roughage in order to remove the dangers of constipation.

The derangement of saccharinous assimilation seemed to Prout to be altogether more easy to acquire, and therefore more common, than that of any other principle because of its low organisation and vital character. Some of this tendency was hereditary or dietary, or a question of temperament, sex and age, or due to exposure to cold and damp, or to malaria. He had been struck by the association of malaria with oxalic acid diathesis in 1832 at about the same time that the correlation with Asiatic cholera was first made (Prout 1848: 19).

The atmospheric or miasmatic theory of disease was widely supported by nineteenth-century physicians who were faced with unsanitary conditions and repeated attacks of small pox and cholera (Pelling 1978). The atmosphere was conceived to be charged with the exhalations from fermentation and organic decompositions; these exhalations were supposed to produce diseases and epidemics. In Prout's extended conception of merorganised bodies, the exhalations were thought of as rapidly rotating particles which were able to penetrate living organisms and interfere with their states of organisation. (A virus analogy immediately springs to mind.) Although a particulate 'germ' theory, it was not the same as the later germ theory of disease since Prout said nothing about contagion. However, like all atmospheric theories of disease it explained the prevalence of disease in marshy areas and undrained unsanitary areas of towns. Although this explanation proved too simple, it at least suggested that a fundamental step in preventive medicine would be sanitary reform.

Any malfunction of this assimilation of albuminous substances was also best discovered from an examination of the urine for an excess or deficiency of urea, the presence of albuminous material, the presence of uric acid and its compounds, or the presence of cystine, xanthine and hippuric acid. All these materials might form, or help form, concretions. Such disorders were influenced by heredity, poor diets, intemperance and worry; but like Liebig later, Prout thought they were more frequently due to malfunction of the processes of secondary assimilation, especially the transformation of gelatine into urea and saccharinous substances. He suspected that malfunction in the production of urea was often a forerunner of diabetes, and hence a very important symptom.

Much less was known about the derangements of oleaginous assimilation, Prout admitted, but clear examples were obesity and leanness, and the malfunctions which produced gallstones. Once more Prout suggested inheritance as a culprit for these derangements. Climate and locality were aggravating causes. He recognised that fat acted as an insulator, even suggesting, contrary to the opinions of Liebig and Dumas, that fats were burned during secondary assimilation to produce animal heat. The other main function of fats seemed to be something to do with nervous and cerebral tissues in which they were associated with large quantities of phosphorus. The quantity of fatty matter in animals appeared to be correlated with their vitality (nervous energy) since leanness was often accompanied by the presence of large amounts of phosphorus in the urine. Prout (1848: 276) hazarded a rather futile prediction that 'the oleaginous principle is much more deeply implicated, not only in the operations of organic life, but in those higher operations connected with the animal functions, than is commonly supposed'. Weakly, he concluded that if this were true then the study of oleaginous pathology would prove of great importance since it might lead to a knowledge of the causes of the most deadly diseases.

The incidental inorganic matters which were also found in foodstuffs or in the products of assimilation, also produced diseases, or could be symptomatic of them. Signs of illness were associated with either their insolubility (as with the magnesium, calcium and ammonium phosphates), or with their solubility (as with soda, potash and ammonia in alkaline urine). Phosphorus was associated with the nerves and nervous action; this probably explained why nervous illnesses were frequently accompanied by an excess of phosphates in the urine. This conclusion was experimentally confirmed by Henry Bence Jones (1845).

It will have been seen from the foregoing that Prout's theory of the nature of matter was intimately related by him to his activities in the field of biochemistry. His theory inspired him to develop an explanation of the major living processes of digestion and assimilation; and these schemes in turn provided him with an explanation of the digestive and urinary diseases in which he specialised as a physician. Prout's scientific career was a complete unity: in the discovery of the ultimate nature of matter lay part of the explanation of life and disease.

Yet both his molecular theory and his theory of digestion and assimilation were almost entirely ignored, and only a few features of his ideas on nutrition and pathology were adopted by his contemporaries. Why was this so? An answer emerges by comparing Prout's methods and ideas with those of his younger rival, Justus Liebig.

Liebig and Prout

It must be said at once that Prout himself was much to blame for the neglect and inefficacy of his metabolic and molecular studies. Despite his unorthodoxy towards Dalton's atomic theory he was inherently a conservative man, and he suffered the chagrin of living to see Liebig and other continental chemists and physiologists build a new science on many of the principles which he had stated or foreshadowed. Had he lived only ten more years he would have suffered the further chagrin of hearing Cannizzaro (1858) disclose the relationship between atomic and molecular weights which he had outlined in the *Bridgewater Treatise*, and the over-confident dismissal of the unitary hypothesis by the skilful Stas (1860).

The brilliant polemical journalist, Thomas Wakley, in a *Lancet* editorial of 1844, made a personal attack on Prout's inertia and conservatism that went directly to the heart of the matter. This must, however, be put in context. The decade 1835–45 brought the German chemist Justus Liebig into international prominence. His tours of England during 1837 and 1842, and his specially invited Reports to the British Association on agriculture and animal chemistry, made him particularly well known in this country; and led to the opening of the Royal College of Chemistry in 1845. Many people hoped in 1844 that Liebig would take over the direction of the College when it was eventually opened, but Liebig recommended instead the appointment of his pupil, A W Hofmann (Roberts 1976). When J F Daniell, Professor of Chemistry at King's College, London, died suddenly in 1845, serious efforts were made to woo Liebig into the vacant Chair. These efforts, made most notably by the President of the Royal Society, Sir Benjamin Brodie, failed because the Governors of King's felt that the Anglican constitution of the College forbade the appointment of a Lutheran professor. The Chair went to Daniell's assistant, William Allen Miller.

Wakley was a reformer who felt that British chemistry had declined since the 1820s and ill-compared with the tremendous improvements and discoveries which had been made by continental chemists in recent years. Liebig's appointment to an English Chair of Chemistry would stimulate the teaching and study of the subject in this country and offset its decline. A patriot asked Wakley why Britain needed a Liebig when it already had a distinguished chemist in Prout. That was just the trouble, argued Wakley:

> Many individuals hold Dr. Prout to be the first of British organic chemists; but several of the doctrines he espouses are opposed to those which are now taught in the continental schools that possess the highest repute.

One had only to read the 1843 edition of Prout's *Stomach and Urinary Diseases* to see this. For instance: Prout had ignored the discovery of pepsin (made in 1836) and relegated the views of Schwann and Müller to a brief footnote; he had old-fashionedly stated that exact compositions of albuminous substances could not be given, and he still regarded them, as well as saccharinous and oleaginous substances, as compounds of just four elements 'merorganised' in the living body of minute portions of certain unspecified 'accidental minerals'; he had completely ignored Mulder's protein; he had more or less ignored Liebig's arguments for the progressive changes of organic compounds in the living state, questioned the accuracy of the analyses upon which such views were based, and ignored the value of chemical formulae and equations in such discussions; finally, Liebig and Prout differed over nutrition and pathology in numerous details, the fundamental difference being that Liebig and Wöhler based all their studies upon the concept of tissue oxidation, whereas Prout never referred to the action of oxygen on the tissues.

This was a formidable list of complaints, all of which, with certain reservations to be mentioned, were justified. Wakley (1844: 489) was in no doubt as to the reason for Prout's failure—his decision not to use chemical formulae, which he scorned as unphilosophical expedients because they did not represent true elementary compositions.

> [We] affirm that had Dr. Prout himself paid more regard to them, we verily believe he would never have placed reliance upon several false analogies, and would have suppressed from the world some hypotheses which are unsubstantiated either by facts or arguments: for notwithstanding that we are told in the preface to the 4th edition on *Stomach and Renal Diseases* that the work is *practical*, and precludes all controversy— we will venture to say that there is no 'standard' work of the present day which contains more hypotheses or more matter of controversy.

This is rhetorical overstatement, for Liebig's *Animal Chemistry* (Holmes 1965) was a far more controversial book than Prout's textbook which, in any case, was addressed to the practising physician and not to the academic chemist. However, Wakley had hit upon an extremely important point, namely Prout's rejection of contemporary chemical notation. This will be examined more closely in a moment.

Finally, for good measure, Wakley felt that Prout had been a baneful influence. Prout deserved his reputation, but as an historic figure, for his work on gastric juice and the alimentary principles. No doubt continental chemists had begun where Prout left off. But why had it been left to them to make the recent advances in animal chemistry? Because 'Dr Prout's name and authority exercises an influence that is detrimental' to the teaching and progress of chemistry in Great Britain. Science always declined when

> the authority of those who, having earned a reputation for themselves, cast unfounded doubts upon the labours of others, neglect and repudiate, without sufficient cause, the methods followed by their competitors, and deny them that honour to which they are justly entitled by their discoveries. We regret to find Dr. Prout in this category [Wakley 1844: 490].

Ignoring the exaggeration of Wakley's remarks, it must be remembered that by the 1840s Prout was almost the sole older practitioner of animal chemistry who was still alive in Great Britain. He was the only important chemist surviving and publishing from the earlier brilliant generations of Davy and Wollaston. Dalton was a dying man, Thomson was old and weary in Scotland, Bostock was to die in a cholera epidemic in 1846 and both he and Hatchett had long ago ceased to interest themselves in chemistry, Ure had entered the field of industrial chemistry, Faraday had become a physicist (a term he never accepted), Fownes's potential was ruined by poor health and Brande had ceased to make original contributions to chemistry. Only Thomas Graham was doing work of any great significance. It was with good reason that Liebig (1837) had exclaimed after his tour of the British Isles 'England is not the land of Science'. English chemistry was at an interregnum. The Chemical Society of London, founded in 1841 by Phillips, Graham and a number of younger chemists who had been fired with enthusiasm for the subject by Liebig's work, and the Royal College of Chemistry finally paved the way for a renaissance of English chemistry in the figures of Graham, Brodie, Frankland, Williamson and Odling (Bud 1980).

Prout made no reply to Wakley's attack, but it is undoubtedly significant that there are indications of a changing attitude in the final edition of Prout's textbook. There his antipathy towards formulae seems to have been waning, protein and pepsin are mentioned, and there are several favourable comments on Liebig's work (Prout 1848: 470, 503). However, Wakley's general criticisms were still valid and the textbook's lack of chemical formulae and definitive account of continental work, led to its rapid replacement by other texts, notably that of Golding Bird (1844).

Why was Liebig so successful compared with Prout? Let us examine Wakley's suggestions, and some others, in more detail. First, there is a purely biographical issue. Prout was a busy physician who was able to devote only his meagre spare time to the investigation of animal chemistry, whereas Liebig was an industrious academic chemist who controlled the activities of a fairly large research school. Moreover, in the 1830s and 1840s Liebig was at the peak of his mental powers, whereas Prout, Liebig's senior by eighteen years, was subjected to a deafness which seems to have profoundly affected his scientific output. It must also be admitted that Prout was by nature of a shy, withdrawn and non-controversial disposition, despite his brush with Wilson Philip in 1831; Liebig, on the other hand, was of an extroverted and polemical mould.

Liebig was not one who dwelt on the merits of his predecessors and he made little reference to Prout's writings. Prout seems to have been slightly shocked by this cavalier treatment from his younger contemporary who arrogantly suggested that physiologists and animal chemists before *him* had never asked the right sort of questions (Liebig 1842b: 131). But Prout (1843: Preface) had been asking Liebig-type questions about the progressive chemical changes that took place in digestion and sanguification ever since 1816 when Liebig was still a schoolboy.

If the factor of personalities is put on one side the problem can be rephrased in terms of the books which contained their ideas on metabolism. Why was Liebig's *Animal Chemistry* so much more influential than Prout's *Bridgewater Treatise* or his *On Stomach and Urinary Diseases?* A general answer to this question has already been suggested, namely that Liebig's book was specifically addressed to the theoretical chemist and physiologist, i.e. to the publishing research scientist. On the other hand, Prout's *Bridgewater Treatise* was presumably intended for the consumption of a lay audience of Christian apologists; and his pathology textbook was addressed to the practising specialist on urinary and digestive diseases who was (according to Prout) not interested in purely theoretical issues. Thus in terms of presentation and publicity Prout never offered his professional colleagues in chemistry and physiology a seriously and well argued discussion of his ideas on matter, nutrition, metabolism and pathology. Liebig did. One perceptive critic saw this clearly and wrote:

> We think that it would not have been amiss to have inserted a few references to Liebig's peculiar opinions, as a guide to the student in the comparison of them with Dr. Prout's views. A good deal of trouble might have been thereby spared to those who ... desire to become fully acquainted with the points at issue between these two distinguished chemists. We trust that Dr. Prout may see the desirableness of publishing, in a separate form, and with more amplification, his opinions on these controverted topics, particularly specifying the evidence on which his own views are founded. Professor Liebig's work is almost entirely of an argumentative kind; the data on which his reasonings are founded are for the most part specified; and thus every reader, possessing a competent knowledge of the subject, can form his own opinion—from his knowledge of the probable truth or error of the data, and from his estimate of the logical precision of the reasoning,—as to the value of the conclusions drawn and set forth by the author. In Dr. Prout's treatise, on the other hand, there is more of assertion, and less of even attempts at proof, the data being, for the most part, locked up in the author's own laboratory; and until Dr. Prout shall see fit to give them to the public, he must be content to have his opinions freely questioned, and the accuracy of his conclusions suspected [Anon 1843: 478].

Just as Prout's molecular theory never received serious attention (except from Henry) because he never published it in paper form, his ideas on metabolism were equally unfruitful. I have not been able to find a single detailed discussion of Prout's ideas on assimilation in either English or continental literature of the period 1830–50. On the other hand, his work up to 1827 was well assimilated both in Great

Britain and on the Continent. It seems, therefore, that as soon as Prout developed his ideas through the medium of his two books, they ceased to be assimilated.

As examples, consider Prout's classification of foodstuffs into saccharinous, albuminous and oleaginous, and his suggestion that milk was the prototype of a balanced diet. Both of these doctrines were adopted by a large number of authorities, including continental ones such as Liebig, Müller and Marchand. These doctrines are to be found in both the *Bridgewater Treatise* and in *On Stomach and Urinary Diseases*, so it might be thought that they owed their wide dissemination in the literature to these two publications; but this is not the case. Significantly, these two conceptions are to be found in the earlier Copley Medal paper of 1827 which went into French, Italian and German translations, and received the commendation of Berzelius (1829). It is probable that Liebig (1842: 42) adopted these ideas from this source. However, Müller (1838: i, 479) in his *Physiologie* and Marchand (1844: 392) in his *Lehrbuch der physiologischen Chemie* show that there was another indirect source of information, namely the English translation of Blumenbach's *Physiology* by Elliotson (1828) which contained extensive notes. Marchand (1844: 290–1, 346–52), as we shall see in the next chapter, was an enthusiastic supporter of Prout's multiple weights hypothesis, and in his biochemistry textbook he lost no opportunity of mentioning and praising Prout's 'magnificent insight'. After citing Prout's classification of foodstuffs, Marchand added that Prout's work on nutrition and digestion was still unpublished (in 1844!), but that John Elliotson had quoted from Prout's unpublished book on the subject.

It would appear then that Marchand did not know of either Prout's *Chemistry* or of his pathology textbook when he compiled his own influential review of animal chemistry. The same ignorance appears to be true of Müller and Liebig, although Liebig was never very careful to cite his reading of his sources of information. The fact that a kindly German disciple of his protyle theory like Marchand knew nothing of Prout's biochemical work after about 1827 has great bearing on the reason why Prout's ideas were not disseminated. The *Bridgewater Treatise* was translated into German even though there was not the same interest in natural theology in Germany as in Great Britain; but no German chemist shows signs of having read it. The German translation of Prout's *On Stomach and Urinary Diseases* in 1843 probably appeared too late for Marchand's use, even if he would have been attracted to it by its clinical title. By 1843, in any case, the revolutionary impact of Liebig's *Animal Chemistry* had altered the subject beyond anything that Prout's two books might have accomplished.

Our original question can now be rephrased for a second time. What was so special about Liebig's *Animal Chemistry* compared with Prout's treatises? 'Few works of comparable scientific import', writes a modern commentator on Liebig's book, 'have provided as many apparent paradoxes as did this remarkable book'. Holmes (1965) has enumerated these paradoxes as follows. Although it laid the foundations for modern biochemistry, few of its details withstood the test of time; although Liebig apparently feared physiologists would not understand his chemical approach, it was largely an elaboration of methods familiar to Prout and the early animal chemists; although Liebig claimed his methods were quantitative and exact, he continually resorted to 'common-sense' evidence to support his speculations; although Liebig introduced biochemical equations for the first time, virtually none of these were correct in detail; despite the success of the first edition of his *Animal Chemistry*, Liebig gave up the attempt to publish an expanded new edition. The same author has suggested that Liebig's success was due to the fact that his approach was

> comprehensive enough to threaten the authority of more orthodox physiological and medical theories. Consequently he was bound to evoke controversies and emotional responses. Only a few years later Liebig's ideas were being judged by the new standards of empirical investigation which his 'Animal Chemistry' itself had helped to foster, but by comparison with which the book seemed to retain too much of the older speculative ways of thought [Holmes 1964: ix]

It cannot be said that Prout gave a comprehensive account of digestion and metabolism, for although paradoxically it was 'all embracing' and supported by a theory of matter, it was insufficiently detailed in its actual presentation. For example, where Prout spoke vaguely of the formation of gelatine from albumin by the loss of carbon dioxide in the tissues, Liebig (1842b: 142) boldly wrote the following equation

$$3(C_{48}N_6H_{36}O_{14}) + 4HO = C_{36}H_{28}O_6 + C_{108}N_{18}H_{84}O_{40}$$
($\tfrac{1}{2}$ atom of choloidic acid) (2 atoms of gelatine, $C_{54}H_{42}N_9O_{20}$).

And whereas Prout had suggested that in destructive secondary assimilation gelatine was broken down into urea, etc, Liebig specifically demonstrated the possible products by means of an equation.

$C_{108}N_{18}H_{84}O_{40} \rightarrow C_{96}N_{12}H_{72}O_{28} + C_{12}N_6H_9O_9 + H_3O_3$
(gelatine   (2 atoms of   (3 atoms of   (3 atoms of
according to   proteine,   allantoine,   water)
Mulder) $C_{48}N_6H_{36}O_{14}$) $C_4N_2H_3O_3$)

$\downarrow$ $\downarrow$

$C_{10}N_4H_4O_6 + C_2N_2H_4O_2 + H_4O_4$
(uric acid)   (urea)   (water)

Neither of these speculative equations proved to be correct, but they seemed to offer the animal chemist a new and better insight into the mysterious workings of metabolism than had ever been offered by Liebig's predecessors such as William Prout. Above all they conveyed a sense of the importance of trying to correlate quantitatively internal chemical changes with net exchanges with the exterior. Thus although Prout (1813a; 1814a) had made pioneering experiments on carbon dioxide output and had developed clinical methods for urine testing, it was left to Liebig to imply the value of the measurement of

CO_2 output : O_2 imput ::: N_2 output in urine : food imput.

Prout and Liebig shared in common a feeling that chemistry was neglected by physiologists who based their study on comparative anatomy to the impairment of research in their subject. The strong words on this subject which Liebig (1842a: 38) made in his *Agricultural Chemistry* were an extraordinary echo of earlier remarks by Prout (1816a; 1831a). Prout agreed with Liebig (1842a: xv) that:

> From the moment that we begin to look earnestly and conscientiously for the true answers to our questions, that we [should] take the trouble, by means of weight and measure, to fix our observations.

But Liebig went on to emphasise that the results of quantitative analysis should be expressed in terms of formulae and equations; and with this Prout could not agree.

In 1834 Prout had been charged with Dalton and others to investigate chemical nomenclature and formulae; a subject which was then a bone of contention among British chemists. The brief majority report that appeared in 1835 supported the continental system of nomenclature and notation; but in a minority comment, Dalton urged his own pictorial system (*Brit. Ass. Rep.* 1835: xxiv, 207). Although Prout evidently went along with the majority, he privately developed a clumsy and unworkable notation (Turner 1835). 'I have purposely omitted the formulae now so much in fashion among chemists', he wrote in the Preface to the fourth edition of his pathology textbook, 'not only because I consider them clumsy and unphilosophical as conventional expedients, but because I am satisfied that very few, if any of them, represent the true constitution of organised substances'. Five years later, in his final words on the subject, he still maintained the basic untrustworthiness of formulae and equations (Prout 1848: Preface). This positivistic extremism was as much a result of his molecular theory as of his feeling that analytical methods were still too imperfect. Just as Davy had scorned the Daltonian atomic theory for its acceptance of the elementary nature of undecompounded bodies, Prout was wary of an organic chemistry which had not investigated the nature of these elementary bodies. The 'multiple relations' between the atomic weights of the elements offered a real clue to their combinations within organised substances, and Prout (who distrusted the atomic weights used by Liebig) was heartened to find Dumas reducing the atomic weight of carbon to a whole number. Chemists had recently

> reluctantly admitted the existence of such relations among the four constituent elements of organized bodies. Another generation, I have no doubt, will recognise and admit the important consequences to which these relations lead [Prout 1843: Preface].

Here Prout was completely deluded. Whatever the reality behind Prout's hypotheses, whether the unitary theory of matter, or the existence of multiple relations between the atomic weights, they turned out, with one exception, to have no bearing on the development or understanding of either organic chemistry or biochemistry. The exception to this was, of course, the factor of atomic weight determinations. Accurate analyses, formulae and equations were only possible through accurate atomic weights; a slight alteration in the atomic weights of carbon, oxygen, hydrogen or nitrogen could make considerable differences in the proposed formulae of such complex molecules as albumin. To that extent Prout was right to be sceptical of Liebig's results. 'I still believe', he wrote in 1848,

> that the true nature of the lithic acid and its compounds is not at present understood; nor moreover can be understood, till the fundamental relations of what are called the atomic weights of hydrogen, carbon, oxygen, azote, are duly taken into account. The same remark applies to most organic compounds [Prout 1848: 541].

It is strange that Prout did not proceed further along the road taken by Laurent and Gerhardt who accepted formulae as 'recipes' or symbols of chemical behaviour, and not as descriptions of real relationships between atoms in a molecule. Even here the adoption of a particular formula in any sequence of reactions depends upon accurate knowledge of atomic weights. But since Prout believed himself in possession of true molecular weights, there does not seem to have been any real hindrance to his use of formulae. I can see no reason to think that had Prout lived on into the 1850s he would not have proceeded to use synoptical formulae.

If the use of formulae was a fundamental difference between Prout and Liebig, several of the latter's ideas were held in common with the former. As many contemporary reviewers of Prout's work pointed out, Liebig's 'metamorphosis of tissues' was the same as Prout's 'secondary assimilation'; and Liebig adopted Prout's classification of foodstuffs and his ideas on milk without modification, except that he went on to assign erroneous specific functions to the three aliments. Thus Liebig (1842b: 96) divided foods into 'plastic' or structural nutrients (nitrogenous) which were responsible for the formation of new tissues, the replacement of degraded tissues and the supply of protein from mammal milk; and 'respiratory' or fuel nutrients (non-nitrogenous) which were oxidised in the tissues to produce animal heat. These exclusive roles had to be modified in the light of increased experimentation, but partly also because, as Prout had argued, the animal body was capable of transforming one class of nutrients into another. But both Prout and Liebig saw clearly that the physiological properties of living organisms were directly related to the chemical composition of their components.

Liebig's rigid functional division of foodstuffs into plastic and respiratory implied that fats and carbohydrates were of equivalent importance, and that either fats or carbohydrates were not absolutely necessary in a diet. This was in contrast to Prout's teaching that all three aliments were absolutely essential. Time has proved Prout correct. Liebig's rigidity also led him to argue that muscular work was obtained from the degradation of nitrogenous tissues (gelatine) into urea and its derivatives. In this case work should lead directly to an increased output of urea and uric acid, and hence 'the amount of tissue metamorphosed in a given time may be measured by the quantity of nitrogen in the urine' (Liebig 1842b: 245). Such was the attractive force of this quantitative proposition that it took several years before it was understood that muscular energy came in fact from non-nitrogenous foods (Liebig's respiratory foods), and that proteins were principally responsible (as Liebig had admittedly suggested) for the growth and maintenance of tissues (McCollum 1957). Prout had said nothing concerning the origin of muscular work, and therefore he missed the opportunity seized, albeit incorrectly, by Liebig for mapping out 'a program for investigating these functions quantitatively under various conditions of nourishment and activity (Holmes 1965: lviii).

Prout (1834a: 525; 1845: 474) also missed the opportunity to say anything concerning the origins of animal heat. Liebig (1842b: 17–36) on the other hand, was primarily concerned to demonstrate how the overall chemical transformations in metabolism might be related to the production of heat and work. This, together with the use of quantitative equations, was the key to Liebig's success. There was one other factor. Both Prout and Liebig constructed metabolic theories that have proved to be far too simple. But there was a greater innate success factor in Liebig's over-simplification compared with Prout's because the former constructed a theory (and a methodology) which was testable, or falsifiable, in all sorts of details. Most of these details proved to be incorrect but this turned out not to matter because, as Holmes (1965: cxv) has perceptively remarked:

> First, [Liebig's] genius and chemical experience enabled him to perceive what must be true of the chemistry of organisms even if technical difficulties had so far prevented physiologists from discovering the phenomena directly. Second, his thorough knowledge of organic compounds, at a time when most physiologists were still relatively unfamiliar with them, enabled him to make an exposition which appeared to them a revelation of phenomena they had not previously understood. Third, a general dissatisfaction with older biological ideas made physiologists receptive to a new approach which promised more than it had yet delivered. Fourth, Liebig's influence as a teacher helped him to attract investigators to explore further his enticing theories. Finally, the rapid spread of experimental physiology provided the means and will to transfer his ideas to the direct investigations of animals.

With the exception of influence as a teacher, Prout shared all these characteristics with Liebig. But since he did not construct 'enticing theories', or rather because he did not provide testable chemical and physiological details within his theories, his ideas lacked impact; and they were without influence both in Britain and on the Continent because Prout was only an individual without much time to devote to a very complex field of experimental and theoretical investigation.

Henry Bence Jones (1850: 126), a pupil of Liebig's and an expert in the field of urine chemistry, succinctly summarised the respective merits of Prout and Liebig as follows:

> Before anyone else without doubt, Dr. Prout previous to the third edition of his work in 1840, had formed for himself decided ideas on these questions (viz, the relationship between urine and metabolism, etc.); but from the terms secondary digestion, vitalization, conversion, reduction, and diathesis, others could not obtain clearness, but only some confused comprehension in which no study of Dr. Prout's work could render perfectly distinct. The ideas of Berzelius [on Animal

Chemistry] were but little known here, and it was not until Professor Liebig's work on Animal Chemistry appeared, when, with deep chemical knowledge and great clearness of ideas, he set forth the relations of the urine to the changes in the solids and fluids effected by use, and the action of the inspired oxygen, that the meaning of the language of Dr. Prout became evident.

7

The Prout Debate 1816–60

> In his anxiety to monopolize the honour of finishing the fabric of Dr. Prout [Thomson] unhappily for his fame, deranged the whole edifice, and instead of fixing the key-stone, has actually laboured unwittingly to displace it [Ure 1825: 126].

In the paper which revealed Prout's identity and clarified his somewhat obscure remarks, Thomson (1816b) seized upon the suggestion that 'the specific gravity of any body [on the air = 1 scale] may be obtained by multiplying the weight of the atom [on O = 1 scale] by half the specific gravity of oxygen gas'. He noted that this relationship indicated that:

$$\text{'Atomic' weight} = 2 \text{ (specific gravity)}$$

and that gaseous substances could be classified into three groups according to the relationship between 'atomic' weight and specific gravity, on the scale O = 1 for both weight and volume. (It should be understood that by 'atomic weights' Thomson included 'compound atomic weights', that is 'molecular weights'.) In the first group, the values were identical; in the second group, the 'atomic' weight was double that of the specific gravity; and in the third group, the 'atomic' weights were four times as great. This was the origin of 'two volume' and 'four volume' chemical formulae which was to befuddle organic chemistry until the mid-1860s (Rocke 1984).

These relationships are so good that it would be difficult not to suspect that there has been an adjustment of figures. In fact, in transforming from specific gravities on the scale air = 1, to O = 1, Thomson manipulated the final decimals occasionally so that there was complete accord with the known molecular weights. However, the differences are so small that it would be extremely unfair to imply that these alterations were unjustified. On this basis, therefore, it

	Specific gravity O = 1	Atomic weight O = 1
Group 1		
Oxygen	1.000	1.000
Ethane	0.876	0.875
Group II		
Phosgene	3.095	6.190
Chlorine	2.250	4.500
Cynogen	1.621	3.242
Carbon dioxide	1.374	2.750
Hydrogen sulphide	1.062	2.124
Sulphur	1.000	2.000
Nitrogen	0.875	1.750
Carbon monoxide	0.875	1.750
Steam	0.5625	1.125
Carbon	0.375	0.750
Hydrogen	0.0625	0.125
Group III		
Hydrogen iodide	3.986	15.944
Hydrogen chloride	1.1557	4.623
Ammonia	0.53125	2.125

seemed to Thomson that the atomic weights of the elements (with the exception of oxygen) were, as Prout had implicitly stated, double their specific gravities, or volumes.

Just as Thomson had earlier taken up the flag of Daltonian atomism, he now took up Prout's integral multiples hypothesis; and with the object of verifying it, he engaged upon an arduous programme of analyses from 1818 to 1825. That *verification* was Thomson's sole aim is made clear from several of his remarks during the course of these researches. The fact that so many atomic weights do fall close to integral values was, of course, an apparent experimental confirmation of this commitment. The few outstanding exceptions to integral weights, like chlorine and copper, could only present a problem to Thomson's followers when analytical techniques had been made more rigorous and exact; even then a way out of this problem had already been suggested by Prout.

In 1817 Thomson published a fifth edition of his textbook in which he included hypothetical atomic weights that supported Prout's integral multiples hypothesis. Thomas Brande (1818), who continually opposed Thomson personally for some real or imagined slight to his Royal Institution colleague, Sir Humphry Davy, attacked him from the pages of the Institution's journal. Bluntly, he accused both Prout and Thomson of rounding-off their experimental values. Thomson's citation of Prout's results from the specific gravities of oxygen, hydrogen and nitrogen in his textbook, without giving the reasoning which lay behind their adoption, led Brande to examine Prout's original papers. After suggesting that Prout begged the question over the chemical constitution of the atmosphere, he suggested that the reasoning of Prout and Thomson was circular.

> It is curious to observe the gross application of the calculations of the two doctors. Dr. Prout deduces the specific gravities of oxygen and azote from the weights of their atoms 10 and 17.5, allowed by Dr. Thomson; and Dr. Thomson calculates the weights of the atoms from the specific gravities determined by Dr. Prout.

Prout's argument from 'theoretical considerations' for the specific gravity of hydrogen was pretentious, Brande continued, since all it amounted to really was the random adjustment of experimental findings. In any case, the accuracy of the ammonia method was suspect.

> To us it appears that the sources of error in calculating from a body composed of three volumes are much more numerous (taking more particularly into consideration the nature of that body) than any which could arise from the old method of direct weighing.

Brande also had sport offering contradictory quotations from Prout and Thomson, but most of his criticisms were only debating points. Finally, he noted the hypothesis of multiple weights, but rejected it as unconvincing with the valid comment that 'we always have a jealousy of these *just* numbers'. The many changes which Thomson had introduced into atomic weight values between 1813 and 1817 were strong additional reasons for caution in the adoption of any more new values.

This review has been mentioned at length since it was one of the few criticisms of Thomson which was also addressed to Prout. In the following year, Thomson (1818) published a new list of atomic weights to three decimal places on the scale, $O = 1$. It contained several whole numbers and several rounded fractions that were one eighth multiples of the atomic weight of hydrogen ($8 \times 0.125 = 1.000$). Although these numbers could be interpreted as deliberate deceptions by Thomson, it is more likely that he believed that experimental errors should not, in the words of William Henry (1829: ii, 663) be 'allowed to countervail a general law of so much simplicity, and supported by so many probabilities'. Surely all atomic weights (including molecular weights) were integral multiples of the atomic weight of hydrogen. 'Nature delights in simplicity', he declared. 'Hence I am led to expect simple numbers for the weights of the atoms of the simple bodies' (Thomson 1820a: 17). In this paper Thomson accepted Prout's theoretical value for the specific gravity of hydrogen, and in the following year he obtained an experimental confirmation of it by the direct weighing of pure hydrogen which he had prepared from sublimed zinc and distilled sulphuric acid. No corrections for the presence of water vapour in the hydrogen and the air were recorded (Thomson 1819; 1820b).

Further new specific gravity measurements and chemical analyses were published by Thomson (1820c; 1821a; 1821b) in the early 1820s and at the same time he took the opportunity to publicly express a wish that Prout would continue with his work on atomic weights.

> Dr. Prout had the merit of discovering that even the experiments of Biot and Arago gave the specific gravity of hydrogen gas above the truth and [had] the sagacity to determine the true specific gravity without any original experiments of his own; but merely by a close and ingenious comparison of the experiments of others. This is a degree of skill that places its possessor in a more elevated rank than a mere experimenter, and induces me to prognosticate with confidence that if Dr. Prout persevere in the career which he has begun with so much ardour, the science of chemistry will be indebted to him for discoveries of a far higher and more important kind than have hitherto been made [Thomson 1820c: 166].

This fine compliment is perhaps the origin of the tradition (Benfey 1952) that Prout was only a theoretician, and not an experimentalist; for Thomson (1820c: 247) appeared to say that Prout, unlike Berzelius who 'trusted entirely to experiment', was completely above experiment. This was an exaggeration of course, but perhaps he was trying to be provocative: chemistry, a qualitative science, needed quantitative minds. Indeed, thought Thomson (1820c: 167), the interest of analytical minds like those of Prout and the young John Herschel in the phenomena of chemistry was of the greatest importance, and they would reap 'a rich harvest of discovery and renown, if they chose to devote to this delightful science the requisite degree of attention, and combine their experimental skill with their mathematical acquirements'.

Remarks such as these were bound to antagonise Berzelius (1912– 36: iii, 58) already growling his disapproval of Thomson in private correspondence. But Thomson (1821a: 242) went even further, and specifically objected to the number of decimal values which Berzelius had given in his list of atomic weights; both on the valid grounds that such accuracy was unattainable, and the invalid one that nature was simple and not give to such complications. Prout's integral multiples laws, Thomson declared, was the 'third great step in our investigation of the atomic theory', and Berzelius had paid the penalty of inaccuracy for ignoring it. His lack of precision was entirely due to the fact that he had no such theoretical criterion to guide his experimental findings. Thomson, on the other hand, believed that he had developed a foolproof and accurate experimental technique for the verification of Prout's hypothesis.

The first major public attack on Thomson's support for the integral multiples hypothesis came not from Berzelius, but from another Glasgow chemist, Andrew Ure, in 1821. By all contemporary accounts Ure was an unpleasant character; but he was a competent chemist who had seen the wisdom of adopting the hydrogen scale of equivalents, and of adopting a single volume of hydrogen as the equivalent of the atom. He had even appeared to favour Prout's protyle hypothesis, which he ascribed to Davy (Ure 1821a). Yet he was no friend to Thomson; in a long savage review of the sixth edition of Thomson's *System*, written for Brande at the Royal Institution, Ure (1821b) deplored the hypothetical character of Thomson's reasoning.

> Whenever he begins to generalise, his technical decision [*sic*] of manner leaves him, and, to the surprise of the readers of those clear [experimental] details, which he had merely transcribed from experimental chemists, he becomes obscure and contradictory. To this defect a more serious fault has been added; and which, progressively gaining force, has of late grown almost intolerable; we mean, the preference of hypothesis to facts on innumerable occasions, so that it is difficult for the experienced chemist, and impossible for the tyro, to distinguish between them in his works.

Ure's review quickly degenerated into personal abuse, but above all he castigated Thomson for the suggestion that chemists had worked without any guiding principle. Thomson's book was filled with errors on every page which were 'occasioned chiefly by *the incessant twisting, stretching, and curtailing of experimental results, to suit some fantastical atomical*' guiding principle (Ure 1821b: 171).

Ure, and his editor Brande, had nothing to say about Prout on this occasion, but they did suggest that Thomson's division of the atomic weight–specific gravity relationship into three classes of substances was a trivial effect due solely to the fact that chemists had adopted a dual standard of oxygen for weights, but air for specific gravities. This is quite true, but without a molecular theory of chemical combination Thomson was surely right to have treated it as a possibly significant series of relationships.

Thomson (1822a) was much too busy with his experiments to reply to Ure and Brande's 'uncandid review' for nearly a year, when he published a long and detailed defence—chiefly of his personal character. Although the review had been written by Ure, Thomson directed his spleen against Brande who he believed had reviewed the fifth edition of this textbook in a similarly vindictive manner. Thomson avoided any reference to the accusation that he had 'adjusted' his experimental results, and instead defended his experimental accuracy on the modest grounds that lack of dexterity had forced him 'to look out for a method in which no dexterity was required'. As a result he had adopted the method of preparing equivalent solutions which, when reacted together, exactly neutralised each other.

His reply occupied thirty five pages; inevitably it was followed a few months later by another twenty-page assault from the Royal Institution which was again a dreary attack upon Thomson's integrity. These personal controversies are arid reading for the historian; yet it is possible that they may have brought Thomson a certain amount of sympathy. If this is the case, then they may also have hindered the serious experimental appraisal of Thomson's work. Some support for Thomson certainly came from the new editor of the *Annals of Philosophy*, Richard Phillips, Thomson having been made Regius Professor of Chemistry at the University of Glasgow in 1818. In 1824, Phillips published a table of equivalent weights which were all integral on the hydrogen scale. But it should also be noted that, although Brande's opposition to Thomson was to prove well founded, Brande (1823) did publish a comprehensive table of integral equivalents ($H = 1$) for the use of his students at the Royal Institution. However, Brande offered no theoretical discussion and, since it seems unlikely that he had suddenly adopted the integral multiples hypothesis, the probable explanation is that, like Davy, he adopted simple numbers for the sake of his students' memories.

At first Thomson had intended to publish his results periodically in the *Annals of Philosophy*, but he soon decided against this in favour of the magnum opus which he published in two volumes in 1825, *An Attempt to Establish the First Principles of Chemistry by Experiment*. He claimed (Thomson 1826: 14) that this book was based upon 'thousands' of accurate experiments which had been made 'in the College laboratory of Glasgow by myself or my pupils under my immediate direction and superintendence'. The *First Principles* was impressively dedicated to the founders and developers of the atomic theory, Dalton, Gay-Lussac, Davy, Berzelius, Wollaston and William Prout. Ostensibly, the book was directed at students of medicine for whom Thomson hoped 'to reduce the whole doctrine of atoms to the utmost degree of simplicity and accuracy', but he warned the reader against bias because his results did not 'exactly coincide with the analytical results of other chemists'. The reason why Berzelius's results were so different, Thomson suggested, was because he had usually worked from the metals themselves, and not from their salts. It was far better to work from neutral salts which could be rigorously purified, claimed Thomson. In the case of zinc, Berzelius, who had worked from the contaminated metal (argued Thomson), obtained the atomic weight 4.03225 ($O = 1$); whereas Thomson, who had worked from purified zinc sulphate, obtained the rounded decimal, 4.25. Unfortunately for Thomson, it turned out that this analysis was incorrect.

Thomson's results were offered as a direct challenge to those of Berzelius whose atomic weights, he declared, were only 'approximate to the truth' and only rarely correct. Berzelius's atomic theory was 'so complicated and intricate, that it would be surprising if it were a true representation of what takes place in nature'. This statement confirms the impression that Thomson's methodological assumption was the principle of economy.

In an historical Introduction based upon an earlier paper. Thomson praised Prout's 'admirable' anonymous article of 1815 'for a degree of sagacity that has seldom been exceeded in chemical investigation, and shows clearly that the author, if he chose, might rise to the highest eminence as a chemical philosopher'. It was to Prout that Thomson (1825a: i, 25) ascribed the observation (which was really his own interpretation on the oxygen scale) that:

> all . . . atomic weights are multiples of the atomic weight of hydrogen; indeed, that all of them are multiples of twice hydrogen, or 0.25, and most of them of 4 hydrogen, or 0.5. He also observed, that in general the specific gravity of the body in the gaseous state, may be obtained by multiplying its atomic weight by 0.5555, or half the specific gravity of oxygen gas; because the oxygen atom is represented by half a volume, but that of most other substances by a whole volume.

He had rapidly convinced himself that Prout was right, and the integral multiple weights law therefore gave analytical chemists a new tool with which to determine atomic weights.

> For every substance, of which I could procure a sufficient quantity to enable me to examine it fully, has been found not only a multiple of the atomic weight of hydrogen, but if we accept [sic] a few compounds into which a single or odd atom of hydrogen enters, they are all multiples of 0.25 or of two atoms of hydrogen.

Thomson altogether avoided any reference to Prout's other hypothesis that the elements were compound bodies made of hydrogen. His only reference to the homogeneity of the elements was made in the context of the eighteenth-century Boscovichean force atom.

> With respect to the notion entertained by Boscovich, that the ultimate atoms of bodies are homogeneous, we are incapable at present of deciding whether it be well or ill founded. It is not likely that any of these ultimate elements has ever come under our inspection. All our simple bodies are most probably compounds. It is possible that the ultimate elements of bodies may be very few—it is even conceivable that they may be reduced to two; but in what way all the variety of bodies with which we are acquainted, could be produced from one single kind of elementary body or atom, I cannot, for my own part, form any conception [Thomson 1825a: i, 31].

This is an interesting statement, for it shows that although Thomson was firmly committed to the principle of the simplicity of nature, and equally firmly that he was in favour of the existence of mathematical relationships between the atomic-molecular weights of chemical substances, he refused to speculate about the unity of matter, or about the candidates for such a theory. If he had agreed with Prout that hydrogen, or hydrogen and oxygen, were primary elements, he would undoubtedly have said so; instead he only went so far as to state that if chemical elements were complex, it was unlikely that any of the 'ultimate elements' from which they were constructed were known to the chemist. Thomson made no deduction that since atomic weights were multiples of the weight of the hydrogen atom, or of the oxygen atom, that one or both might be ultimate elements. His attitude is quite different from that of Prout's, and it once again points to the necessity of dichotomising the so-called 'Prout's hypothesis'. Thomson's position was not unlike that of Mendeleeff and Lothar Meyer fifty years later when faced with the implication that the relationship between the atomic weights described by their periodic law was an indication of the unity of matter (Spronsen 1969). (But whereas Thomson would, like Lothar Meyer, have accepted the implication, but avoided speculation, Mendeleeff actually rejected it completely.)

The great bulk of Thomson's book was concerned with the experimental determinations and calculations of the specific gravities and atomic weights of all the known elements. The careful reader will find a certain amount of evidence that Thomson (1825a: i, 160) adjusted a few results to conform with the integral multiple weights law. For example:

> 4.2519 is the [experimentally determined] weight of an atom of fluoboric acid. The law of Dr. Prout, which will be found to hold in the atomic weights of all bodies, shows us that this number is a very little too high. The true atomic weight is undoubtedly 4.25.

By means of such adjustments Thomson was able to uncover a number of remarkable mathematical relations between the various atomic and molecular weights of substances. Although none of these relations is really significant, they show what Thomson meant and understood by the 'First Principles of Chemistry', and clarify the meaning of his belief that a mathematical or quantitative basis lay behind the qualitative properties of substances. The relations were presented in the form of a series of generalisations in the final chapter.

1. Hydrogen was the lightest known element with an atomic weight of 0.125 (O = 1).
2. The atomic weights of all other elements were multiples of 0.25, or two atoms of hydrogen. This meant that all atomic weights were either whole numbers or multiples of the quarter decimals, 0.25, 0.5, and 0.75. Thomson did not draw any conclusion from these relations that hydrogen was the prime matter.
3. Of the 117 atomic (and molecular) weights which he had determined, only 5 were multiples of a single hydrogen atom; 37 were multiples of two hydrogen atoms (0.25: 11 acids, 11 bases, and 15 elements); 25 were multiples of four atoms of hydrogen (0.5: 6 acids, 8 bases, 2 combustion supporters, 3 combustibles that produced acids, and 6 that were alkalifiable); 50 were multiples of one atom of oxygen, that is integral weights, and 18 of these were elements.

It will be observed that in this third generalisation, Thomson introduced oxygen as a unit of comparison as well as hydrogen. This was continued in further generalisations until the following broad principle was reached.

13. There are five or six of the simple bodies which we have found to combine both with 1 atom and $1\frac{1}{2}$ atoms of oxygen. . . . I have sometimes thought that the anomaly might be obviated by admitting that oxygen in reality has an atomic weight amounting to 0.5 instead of 1 [Thomson 1825a: ii, 463].

Prout had probably already taken this step by 1825, for Thomson's proposition leads to a molecular theory of matter. If the oxygen atom were 0.5, so that two atoms formed a molecular unit of oxygen then, as Thomson pointed out,

not only those simple bodies whose atomic weights are whole numbers, but those likewise, whose weights end in 0.5 are multiples of oxygen. These constitute the whole of the supporters of combustion, and all the acidifiables and intermediate combustibles, except three: namely, carbon, arsenic, and tungsten. Of the alkalifiable combustibles, amounting to 29, 17 would be multiples of oxygen, leaving altogether,

15 simple bodies which are multiples of 2 atoms of hydrogen. Thus 31 simple bodies would be multiples of oxygen, and 15 multiples of 2 atoms of hydrogen.

From this statement it is clear that Thomson hoped to discover rules of affinity from such numerical principles—principles which were the real basis for chemists' gross classifications of substances by their chemical properties. However, it would be difficult to avoid concluding from this analysis that the physical basis for these relations and chemical affinities, was that matter was made from hydrogen and oxygen.

In a final burst of Pythagoreanism, which clearly foreshadows the deductions of Dumas to be considered later, and the transmutation speculations of some less well known chemists, Thomson pointed out some of the extraordinary arithmetical relations between the elements. For example:

$$\text{carbon} = 0.75$$
$$\text{phosphorus} = 0.75 \times 2 = 1.5$$
$$\text{sodium} = 1.5 \times 2 = 3.0$$

And if one made the explicit assumption that the elements were homogeneous, then:

$$\text{phosphorus} = 2 \text{ carbon}$$
$$\text{sodium} = 4 \text{ carbon}$$
$$\text{molybdenum} = 8 \text{ carbon} \qquad \text{etc.}$$

However, Thomson was not explicit concerning this assumption; he merely noted down these relationships without commenting on their physical cause. Yet although he avoided any reference to Prout's protyle hypothesis, it would have been a naive reader who was unable to read into this numerology what Prout had already hinted at, and what Dumas was to make explicit—namely, support for the idea that chemical elements were compound. Despite my earlier qualification then, it appears that like Miers and Dumas, Thomson wondered whether the ultimate units of chemical phenomena were hydrogen and oxygen. But he never made any explicit reference to this possibility.

What sort of reception was given to this extraordinary book? Apart from the inevitable disapproval of Brande and Ure, Thomson received a good English press. Davy, for example, in presenting Dalton with a Royal Medal in 1826, was able to praise Dalton, Prout, Berzelius and Thomson in the same breath (Davy 1839–40: vii, 96–7). Unfortunately, there is no record of Prout's impressions, but it seems likely that while he approved of Thomson's general purpose and was delighted with his demonstration of the law of multiples, he had some reservations about many of the analyses. There is no doubt that the opinion of the majority of British chemists was voiced by Phillips (1825b) in his review of the *First Principles*. Thomson's method, he wrote,

> appears to us liable to exception in very few cases; the work must form a part of every chemical library, and will be referred to as a standard by those who wish to acquire information as to the atomic weights of bodies, or to a knowledge of the experimental means of ascertaining them.

Thomson's book did become a standard among British chemists; Prout was able to say at the beginning of 1827 that his law of integral multiples appeared to have become generally adopted by chemists (Prout 1827: 355). The one exception noted by Prout was Andrew Ure (1825) who had written a polemical review of Thomson's book for the Royal Institution's Journal.

Ure had disagreed fundamentally with Thomson's whole approach. The real aim of chemistry was not, as Thomson seemed to think, to determine the relative weights of substances, but to determine the properties and qualitative relations and affinities between elements and compounds. Experimentally, Ure (1825: 115) thought that Thomson's method of precipitation left much to be desired, 'and hence he has often presented us with results, tallying well with the atomic theory, and with Berzelius [*sic*], which he states as his own, though they could never have been derived from his narrated experiments'. All this followed because Thomson had tried to prove the validity of Prout's integral multiple weights hypothesis which Ure admitted was an interesting and plausible idea. The opportunity for sarcasm was too good to miss.

> Thomson sublimes ordinary zinc in an earthern retort, dissolves a given weight of it in nitric acid, and then expels the acid, by heating the nitrate to redness in a green glass retort. In this way, he obtains at once, to the minutest fraction, every thing which Dr. Prout's atomic multiples require, viz., 5.25 grams of oxide, from 4.25 grams of metal. This felicity of coincidence between his experiments and his theoretical aim is so usual with Dr. Thomson, and with him alone, as to excite no surprise in our minds.

Thomson, although he could hardly have expected favourable treatment from Ure, made an immediate reply. Apart from some well hammered abuse of Ure for his continuous campaign against him he made an honest and serious attempt to defend his belief in quantitative research.

> The whole of chemistry, so far as it is entitled to the name of science consists in the accurate measurement of quantities [Thomson 1826].

Events were to prove Thomson right even though the detailed attempt made by him was to prove worthless.

Ure's 'severe animadversions' on Thomson's *First Principles* were sympathetically noted by Phillips (1826) who redressed the balance of criticism by printing an extract from Benjamin Silliman's glowing and uncritical American review:

> There is nothing, the offspring of the present age, which, so far as we are informed, surpasses this *Attempt to Establish the First Principles of Chemistry by Experiment*. The vast amount of labour performed—the patient and persevering repetition of tedious and often difficult processes, frequently to the eighth or tenth time—the consummate skill discovered in devising and executing the experiments, and the surprising coincidence of the results of analysis with the deductions of theory, excite our astonishment, and prove beyond a question that chemistry, if not founded on intuitive, is built on demonstrative truth.

However, the reception of the *First Principles* was not confined to either fulsome praise or rhetorical abuse; far more serious for Thomson's reputation as a highly credited, or highly discredited, chemist was the attack made on him by a friend, the Glaswegian physician Harry Rainy. Rainy's criticisms, which were made on experimental grounds, were similar to, and continued those made by an Irish physician, James Apjohn, in 1822; and they foreshadowed the searching experimental appraisal of Thomson's work that was published by Edward Turner in 1833.

Rainy (1825) argued that Thomson had made a bad error by underestimating the quantity of water vapour in his hydrogen sample. Rainy had made his own careful experiments and discovered that the Prout–Thomson value for the specific gravity of dry hydrogen was incorrect for any given temperature and pressure; he proposed a vapour correction factor which had the significant effect of reducing the specific gravity of hydrogen from Prout's and Thomson's 0.0694 to 0.0673. This gave an H:O ratio of 1:16:54 instead of 1:16.

Although Rainy's H:O ratio is not correct, since it was based upon Thomson's erroneous value for the specific gravity of oxygen, his result was of great importance since it led him to discuss Prout's integral weights hypothesis. In this first paper, he suggested that the truth of the hypothesis was an open question that was dependent upon the experimentalist's skill.

> If Dr. Thomson's experiment is correct (and of this we can scarcely doubt from the care and attention with which it is performed), it

disproves the hypothesis that the specific gravities of all the gases are multiples by integer numbers of the specific gravity of hydrogen. It is true that 16.54 does not differ from 16 by more than about 1/32 of the whole, and that a very slight change in the number adopted for the specific gravity of hydrogen would account for the difference; but this merely shows how difficult it is to make any experiment sufficiently accurate to decide on the truth of the hypothesis.

Here Rainy hit the nail on the head, and Thomson seemed to have been placed in the awkward position of either admitting that his experiments were at fault, or that his experiments were right, but that Prout's hypothesis was wrong. However, Thomson (1825b) was unperturbed. He replied to Rainy, in a friendly fashion, that he was certain that his own experimental results with hydrogen were all quite accurate. He admitted that there had been an error over the correction for the presence of water vapour, but this error had not been in the direction claimed by Rainy. Far from underestimating the amount of water vapour in hydrogen, Thomson believed that he had overestimated it. He agreed with Rainy's formula for the conversion of the specific gravity of a moist gas into a dry one, and used his own water vapour pressure data to obtain a specific gravity ratio of H:O::1.0036:16, or 1:15.941, which he thought was close enough to 1:16. However, Thomson had not been altogether satisfied by this, and in order to overcome the objections raised by the uncertainties involved in correcting the specific gravities of gases in the moist state for the presence of vapour, he had made an entirely new set of experiments in which he weighed *dry* hydrogen. The drying procedure was very thorough: pure hydrogen was passed through about 37 inches of calcium chloride tubing, and the amount of water absorbed was used as an additional check on the calculations. The results of these measurements were a new triumph for Thomson, even though the H:O ratio proved to be 1.0077:16, or 1:15.87. Nevertheless, he claimed, within the experimental errors, this ratio was sufficiently close to the integral ratio, 1:16.

Rainy (1826) thought otherwise. Far from clearing away his doubts, Thomson's new measurements only strengthened them. In March of the following year he replied more forcefully and critically than in his first paper that Thomson had made errors not only in his old experimental routine with moist hydrogen, but also in the new direct determination of the specific gravity of dry hydrogen. On the basis of some of his own experiments on vapour pressure whose results he found to be different from the published tables of Dalton and Ure, he once more rejected Thomson's old, but amended, H:O ratio of 1:15.941. As for the new determination with dry hydrogen, he thought that it had a number of technically unsound features; but more significant from our point of view, he questioned whether (even on the assumption that the ratio was correct) Thomson's interpretation that the difference from H:O::1:16 of 0.12 in the new ratio H:O::1:15.87 was immaterial.

> The deviation from an integer is 0.12, or about 1/8, which seems small, but we must remember that the utmost possible deviation is 0.5, for were the experimental results to vary from 16 by more than 0.5, it must approach some other integer. The variation from the theoretic result in this experiment, is actually about 1/4 of the greatest possible variation. Dr. Thomson would probably reply, that the greatest possible variation would be 1 and not 0.5, if we compare the atomic weights of oxygen and hydrogen, instead of their specific gravities; but even admitting this, the deviation from the theory is still about 1/8 of the greatest possible deviation. All numbers which have large ratios to any given number must of course nearly coincide with integral multiples of that number.

The larger the ratio the more inevitable became the coincidence that the ratio should reduce to whole numbers. Since the atomic weight of hydrogen was so small, and since it would be taken as the relative weight (H = 1) of a system, then 'the atomic weights of other bodies must nearly be integers'. Surely Thomson would not have so readily admitted the integral multiple weights hypothesis if the basis for comparison had been chlorine, 36.

Rainy's attitude foreshadowed that taken by Stas; atomic weights may approach integral numbers, but this approach need not have any deeper significance. His attitude perhaps also reflects that of the multi-element school that the integral weights hypothesis was bound up with a particular and simple world picture. To Rainy's mind the only satisfactory test of the integral multiple weights hypothesis would be to estimate the weights of atoms which did 'not bear such a large proportion to that of hydrogen'. If it could be demonstrated that the atomic weights of carbon, oxygen and nitrogen were integers, then and only then, would there be 'strong reasons for believing that the other atoms are also integers' (Rainy 1826: 191). This was a sound and sensible empirical attitude. Unfortunately, although it sounded simple enough, the experiments demanded by it were not easy to design, and when they were made the results were not so easy to interpret as Rainy had hoped.

In the meantime, before these experiments were made, Thomson's results were the best available to chemists, thought Rainy. But he emphasised that using them was not the same thing as believing them to be true. Indeed, he specifically warned Thomson against the delusion that because British chemists were apparently adopting his values for the specific gravities of hydrogen and oxygen, this was an absolute belief.

It will have been seen that Thomson got himself into a contradictory position with Rainy. In 1822 he had denied that water vapour made any material difference to the specific gravity of hydrogen; yet in 1826 he conceded an influence. Because of this concession Rainy believed that Thomson should also concede that his experimental results were hostile to Prout's multiple weights hypothesis, and he advised that:

> The subject evidently requires further elucidation and I am sure Dr. Thomson is too candid to object to its being fully discussed, however clear the evidence may appear to his own mind.

In the event, however, the further elucidation came from another chemist who was then studying in Scotland, Edward Turner.

Rainy's achievement was to show that Thomson's specific gravity methods were open to scrutiny and doubt, and that he had by no means established the correctness of the integral weights hypothesis which he had set out to prove. Rainy's case would have been stronger if he had also tackled the value of the specific gravity of oxygen. Like Andrew Ure, Rainy (1826: 193) also came out explicitly and accused Thomson of some fabrication.

> Dr. Prout proposed his views, as a probable conjecture, not inconsistent with the established facts of chemistry, and therefore, deserving of fuller inquiry. Dr. Thomson adopted them as completely proved, and seems to think that they must be admitted to be true, if they cannot be proved to be false. He accordingly introduced them into his *System*, and did not scruple in any instance to modify experimental results so as to correspond with them.

Since Thomson's *System of Chemistry* was such a widely read textbook, Rainy thought that Thomson had done a grave disservice to chemistry by publishing experimental results in the *First Principles* that had been 'vitiated by modifying them to suit an hypothesis, which, whether it be true or false, will tend to retard the science, if adopted prematurely and on insufficient evidence'. Yet, just like Turner only a few years later, Rainy placed on record his cautious opinion that nothing he had written disproved Prout's integral weights hypothesis. It was an hypothesis still open to experimental inquiry.

To this Thomson made no public reply, and Rainy does not seem to have made any further contributions to the debate, or indeed to the science of chemistry. Although there is some evidence that Prout himself began barometric experiments at this time with a view to improving methods of specific gravity measurement (Brock 1970) the next appraisal of the hypothesis was to be concerned with Thomson's other techniques, those of quantitative analysis.

It has been suggested (Rowe 1955: 59) that the full force of Rainy's attack was weakened by the respect and friendly admiration which he felt for Thomson, and the fact that Thomson's tremendous propaganda for the adoption of Daltonian atomism in Great Britain had been accomplished only at the cost that Berzelius's work was less well known. I think that this is only partly correct. Rainy's criticism was as well mannered as Turner's was to be, and all the more potentially effective for being so compared with the gross polemics of Andrew Ure and, to some extent, Berzelius. Rainy was ineffective, however, because unlike Turner he left it for others to decide on the truth of Prout's hypothesis. It must also be noted that throughout the period 1815 to 1825 Berzelius's work was fairly well publicised in Great Britain, largely through the efforts of Thomson, and later Richard Phillips. Through the strength of his own journalism, however, Thomson managed to persuade a large section of British chemists to follow his own interpretation of the atomic theory, and he seems to have shared the suspicions of the Royal Institution clique that Berzelius's work was a mystification rather than a clarification of chemical phenomena. In this respect it is undoubtedly significant that Berzelius's *Essai* (1819), which clarified his intentions, was never translated into English. On the other hand, it must be emphasised that Berzelius's values for the atomic weights were well known to British workers. In the same year that Thomson published the *First Principles*, Davy's friend J G Children (1825) published 'a summary view of the atomic theory according to the hypothesis adopted by M. Berzelius', and Phillips (1825) printed a list of Berzelius's *non-integral* atomic weights transcribed from the *Essai*. In the space of a few months, therefore, readers of the *Annals of Philosophy* (from 1826 it was absorbed by the *Philosophical Magazine*) were able to compare the atomic weights of Phillips, Berzelius and Thomson. It remains a fact, however, that British textbook writers of this decade always seem to have preferred the weights of Thomson or Brande to those of Berzelius.

In 1826 Berzelius published a new series of atomic weights based upon the new criteria of specific heats and isomorphism as well as his old empirical rules (Partington 1964: iv, 166, 212). No integral values were recorded by him and he made no reference to the multiples hypothesis. We know his opinion of it, however, from the devastating reference to Thomson's *First Principles* which he made in his *Jahres Bericht* for 1826, published in 1827.

> This work belongs to those few productions from which science will derive no advantages whatsoever. Much of the experimental part, even of the fundamental experiments, appears to have been made at the writing desk; and the greatest civility which his contemporaries can show its author, is to forget that it was ever published [Phillips 1828].

Berzelius's famous accusation of quackery was grossly unfair; as Phillips remarked, it went beyond the bounds of just criticism. Berzelius was wrong to 'arraign the character of an individual, who may be actuated by motives and principles as pure as his own'. Thomson was an honest and sincere man, said Phillips, and if any deception had taken place, then Thomson had deceived himself more than anyone else.

> It is possible that, misled by a favourite hypothesis, he may, like many before him, have been too eager in seizing facts favourable to his views, and too tardy in perceiving those that are unfavourable [Phillips 1828: 453].

Despite Phillips' defence, by 1828 Thomson's supporters had begun to have doubts. Berzelius (1827) had argued in the German edition of his chemistry textbook that since Thomson's analysis of the reagent barium chloride was inaccurate, and the resultant value of the atomic weight of barium quite erroneous, most of his other atomic weight estimations were compromised. Even Phillips admitted that it appeared as if something had gone wrong; it was Thomson's duty to chemistry to defend his reputation as an analyst, or admit and rectify any mistakes that he might have made.

Unfortunately Thomson (1829) never chose to make any explicit retraction. Instead he replied in kind to the 'foul aspersions of the Stockholm Professor', defended his integrity, and completely avoided the question of the accuracy of his results. He revealed that in 1825 he had sent a copy of his book to Berzelius (1912–36: iii, 58–60) who had replied by letter that Thomson had made an elementary error with the precipitation of zinc carbonate in one of his first experiments. Berzelius had actually said a good deal more, namely that Thomson had altered good experiments (Berzelius's) to make them agree with bad ones (Thomson's) and that

> I cannot avoid telling you that I have not found cause to be happy with your work, which does not give me much confidence in either your experiments or your calculations [Berzelius 1912–36: iii, 58].

And to his close friend Dulong, Berzelius (1912–36: ii, 70) confided:

> I wanted to assure myself about the supposed multiple weights of bodies compared with hydrogen; but the matter is enormously difficult, and at present most of the evidence is against it. I have repeated the principal experiments of Thomson; what a charlatan! None of his results is exact; when working with the atomic weights he prescribes to make a precipitation one obtains filtrates which, after separation, produce further precipitate with the solution employed.

Berzelius seems to have clearly appreciated before Turner that the basic source of Thomson's errors lay in his technique of precipitation.

In defence against Berzelius's remarks which had received such publicity from Phillips' translation, Thomson (1829: 220) foolishly accused Berzelius of altering his own atomic weights of 1819 in the light of the *First Principles*, and publishing them as his own in 1826.

> I am uncharitable enough to believe, that it was in order to prevent his countrymen and the Germans from being aware of the benefit he derived from my labours, that this attack upon me was made. I had touched his selfish feelings, and disturbed those dreams of chemical sovereignty in which he had been evidently indulging.

With this preposterous charge, Thomson remained steadfast in his belief in the truth of Prout's integral weights hypothesis and in the accuracy of his own atomic weights. The only changes he was prepared to make were all consistent with Prout's hypothesis; and he dismissed Berzelius's non-integral value of the atomic weight of barium.

The analysis of oxalic acid proved particularly troublesome. In the *First Principles* Thomson (1825a: ii, 100) had deduced that oxalic acid crystals contained over half their weight of water, but just before publication, Prout 'of whose accuracy and information I entertain a very high opinion', had written to him stating that he found much less water in the crystals than Thomson. The latter repeated his analysis, but obtained the same result. After the publication of his book, Prout wrote again to Thomson (1829) to say that he also stood by his analysis. This analytical discrepancy between Thomson and Prout shows the great difficulty the former experienced in completely removing water of crystallisation from the salts which he analysed. It also suggests that Prout may have had some reservations concerning Thomson's other analyses before the critical onslaughts of Berzelius and Turner were published. Later, Thomson (1831: ii, 15) said Prout was right.

Thomson's reputation never completely recovered from Berzelius's attack, for although his work had been founded on experiment and not done at the writing desk, except in the sense that he had preconceived ideas about the kind of results he should obtain, and he was honest enough to admit this whereas other scientists of the period took trouble to disguise their precommitments, it became apparent in 1828 that many of his students' experiments were inaccurate, or based upon false assumptions. As J W Mallet (1901) pointed out, from the incorrect assumption that air was a chemical compound, Thomson deduced incorrect densities for hydrogen and oxygen and chlorine, which led to incorrect values for the combining weights of his principal analytical reagents, hydrochloric acid and barium chloride. Since the majority of his determinations hinged upon the precipitation of barium sulphate until the solute was neutral to barium chloride, numerical error was introduced into all estimations that were based upon the use of barium chloride. In addition, and this was not known by Thomson, or even by Berzelius, the heavy precipitate of barium sulphate always carried down with it some of the soluble reagents and the solute, thus rendering useless all Thomson's analytical care. The method of precipitation was suspect in any case since, as Berzelius pointed out, none of the preliminary experiments, in which the equivalent quantities of two pure salts were estimated, were recorded in the *First Principles*. Mallet concluded that:

> Thomson must have been satisfied with only very moderate accordance between the figures of individual experiments and have accepted the mean of these as exact, without thought for the large probable error involved—then, reducing his figures to correspond with the standard assumed, namely oxygen as unity, and expressing the results in *grains* he had but small quantities to work with in the final experiments which he does report, and the absolute errors are seemingly but very small.

It is ironic that, because of an internal cancellation of inaccuracies, Thomson's values are in many cases better than those of Berzelius when compared with modern values (Partington 1964: iv, 226).

In tracing the fortunes of Prout's hypotheses through the work of Thomson, Prout has been overshadowed. This is historically correct, for until 1831, Prout took no part in the controversy over his work. The reader must therefore be patient if Prout is once again placed in the background while the work of his friend Edward Turner is briefly reviewed.

The work of Edward Turner

Edward Turner (1796–1837) studied medicine at Edinburgh University where he later lectured on chemistry before he became the first Professor of Chemistry at the new University of London in 1828

(Terrey 1937). He began his career as a disciple of Thomson and went on record that he recommended:

> the careful study of his recent admirable treaties on the *First Principles of Chemistry* to every one who feels an interest in the science; nor [can I fail to express] my admiration of the profound sagacity, unwearied industry, and great experimental talent displayed throughout the whole of that work [Turner 1825: vi].

He made no reference to Prout's protyle hypothesis; but he appears to have been convinced of the multiple weights hypothesis, while positivist towards the hypothesis of atoms. The essay which contained these remarks was expanded in 1827 into one of the best of nineteenth-century textbooks, the *Elements of Chemistry*. This book was even acclaimed, despite its allegiance to 'Apollo Magnus' Thomson, by Brande (1827) at the Royal Institution.

Before his translation to London, Turner (1828) was led through his mineralogical interests to investigate the atomic weight of manganese which Thomson had estimated by the precipitation reaction between manganese sulphate and barium chloride. Now Turner had seen Berzelius's criticism of Thomson's use of barium chloride, and his correction of the atomic weight of the reagent, so out of concern to make his own experiments with manganese as accurate as possible, he had quickly repeated Berzelius's analysis and found it confirmed. Disconcerted, he drew Thomson's attention to it and urged him to check his experiments 'without delay; since an error in the atomic weight of barium will at once vitiate an extensive series of his most elaborate analyses'. As we have already seen, Thomson (1829) replied that he saw no reason to doubt the accuracy of his analysis.

Turner gave little sign in 1828 that he would devote the next five years of his life to the redetermination of the most important atomic weights. But he undoubtedly made this decision in view of the large discrepancies between the atomic weights of Thomson and Berzelius. In order to make certain that the task was worthwhile he spent some time on an accurate analysis of barium chloride; but because of his teaching appointment in London, he was unable to publish the result until 1829.

In this important paper, Turner (1829) rejected Thomson's claim that when 106 parts of barium chloride solution (70 Ba and 36 Cl) were mixed with 88 parts of potassium sulphate solution, a perfect neutralisation occurred so that no barium chloride or potassium sulphate could be detected in the filtrate; he agreed instead with Berzelius that two per cent of the barium chloride was to be found in the filtrate. Thomson had made two mistakes, suggested Turner. First, he had probably used an impure sample of barium chloride; second, he and all analysts, including Berzelius, had been unaware that when the two salts were mixed together, some potassium sulphate clung to the barium sulphate precipitate and caused an error in the barium estimation. By using specially prepared barium chloride, and carefully washing the precipitate, Turner found in favour of Berzelius's analysis.

	Thomson	Berzelius	Turner
Ba	66.037	65.926	65.984
Cl	33.963	34.074	34.016

Turner carefully abstained from any conclusion concerning the atomic weight of barium until that of chlorine (upon which it depended) could be checked. (Thomson's and Berzelius's values for the atomic weights of Ba and Cl from these figures were 70 and 36, and 68.726 and 35.430 respectively.) It had become clear to him that a good deal of careful experimental work would be necessary in order to verify or refute Prout's integral weights hypothesis. Consequently, in the third edition of his textbook, published in 1831, he retained Thomson's values for the atomic weights and explained to Berzelius, who had thought this an inconsistent policy, that:

> At a distance, you cannot fully appreciate my position. Many here are prejudiced in favour of Thomson's equivalents merely in consequence of their *simplicity*, and I saw clearly that I had no chance of producing conviction in others until my experiments had been considerably extended [Berzelius 1912–26: iii, 284].

Turner (Christison 1837) had placed himself in the delicate position of 'umpire between two of the greatest of living chemists'!

Meanwhile, a provincial chemist from the Plymouth Institution, John Prideaux (1830), who had planned to publish a table of equivalents based upon Thomson's values, had grown tired of waiting for Turner's promised reappraisal; he therefore offered his own experimental check upon Thomson in 1830. Prideaux found that barium chloride and zinc sulphate neutralised one another in exactly the proportions stated by Thomson, and he therefore expressed confidence in the remainder of Thomson's values. He probably obtained similar results to Thomson because of a defective drying procedure, and he clearly had not appreciated Turner's warning concerning the precipitation method. However, even Prideaux expressed doubts concerning Thomson's value for the atomic weight of nitrogen, but modestly concluded that 'whilst the ablest chemists are at variance on such simple experiments, we must be content with approximations'. It is not surprising to find that Prideaux was willing to adopt a compromise between Thomson and Berzelius; he suggested that the sensible procedure would not be to dismiss Thomson's concise and facile numbers, but to average them with those of Berzelius.

	Thomson (O = 1)	Berzelius (O = 100)	Prideaux (O = 1)
Al	1.25	171.167	1.2
As	4.75	470.042	4.73
N	1.75	177.036	1.76
Ba	8.75	856.88	8.66
C	0.75	76.437	0.76
Cl	4.5	442.65	4.46
O	1.0	100.00	1.0
Ag	13.75	1351.6	13.63

These average figures entail an abandonment of any attempt to reveal a law of integral multiples.

In September 1831 the first meeting of the British Association for the Advancement of Science was held at York (Morrell and Thackray 1981) and an *ad hoc* Chemistry Committee resolved that it was of the utmost importance that

> Chemists should be enabled, by the most accurate experiments, to agree in the relative weights of the several elements, Hydrogen, Oxygen, and Azote, or what amounts to the same thing, that the specific gravity of the three gases should be ascertained in such a way as would insure the reasonable assent of all competent and unprejudiced judges [*Brit. Ass. Rep.* 1832: 53].

In the following June, at the meeting held in Oxford, Turner (1832) presented a report of his investigations of the atomic weights of lead, silver, chlorine and bromine which he published in full in the *Philosophical Magazine*. Since the adoption of the integral multiple weights hypothesis had taken nearly all its proof from Thomson's *First Principles*, Turner began 'I turned to that work with the view of putting some of the statements, contained in it, to the test of careful experiment'. He reminded his audience that he had shown that the king-pin of Thomson's analytical structure, the analysis of barium chloride, was in material error. Although Thomson had not acknowledged his error in the reply to Berzelius in 1829, Turner was gratified to note that Thomson had quitely lowered the atomic weight of barium from 70 to 68, in a new edition of his *System* published in 1831. In this, although he avoided any discussion of the integral weights hypothesis (perhaps because of Turner's investigations) Thomson's atomic weights remained integral on the hydrogen scale.

In his 1832 paper Turner gave only a few of his results, and the full experimental details were reserved for the ears of the Royal Society in the following year. He drew three conclusions from his experiments:

1. The atomic weights commonly used by British chemists have been adopted without due inquiry, and several of the most important ones are erroneous.
2. The hypothesis, that all equivalents are multiples by a whole number of the equivalent of hydrogen, is inconsistent with the present state of chemical knowledge, being at variance with experiment.
3. The subjoined equivalents are very nearly correct:
   Lead   103.5   Barium   68.7   Nitrogen 14
   Silver   108   Chlorine 35.45

As a result of this Oxford paper, the Chemistry Committee of the British Association urged Turner to extend his work on atomic weights. Through references by Berzelius, Stas and others to Turner's researches, the impression has grown up that this work was sponsored by the British Association from the very beginning. As can be seen from the foregoing account, however, this sponsorship came only in 1832 when Turner's work was approaching completion; after 1833 illness prevented him from publishing anything further on the subject.

At the same Oxford meeting in 1832, Prout described his measurements of the specific gravity of air, and the Chemistry Committee (of which he was a member) asked him, together with John Dalton, to make fresh experimental studies of the specific gravities of oxygen, hydrogen and carbon dioxide. The initiative for these experiments came from Dalton, as may be gathered from a previously unpublished letter from Daubeny (1831) to Prout, 27 October 1831.

> I wish you would agree [??] to meet at Oxford next July, when we propose receiving the British Association lately organized at York. Dalton wants much to have the specific gravity of Hydrogen and other fundamental points in the Atomic Theory settled by a sort of Chemical Committee, and as he promises to be of the party at Oxford next year it strikes me that a better time could not be chosen for such an undertaking. My laboratory, such as it is, could be placed at your service, though I had rather leave the decision of a point requiring such delicate manipulation to more skilful hands.

There is no evidence, however, that the intended collaboration between Prout and Dalton ever took place; the results of the later British Association award of £40 to Prout and Thomas Clark of Aberdeen for the specific gravity measurements in 1839 are equally hidden.

This flurry of interest in atomic weights, specific gravities and the multiple weights hypothesis was also reflected in the masterful *Report on the Present State of Chemistry* made to the Association by James F W Johnston (1832), a pupil of Thomson's who had also studied with Berzelius. In his report, Johnston mentioned the disagreement and confusion among chemists concerning the relation between atomic weights and volume weights. According to the molecular theories of Ampère and Dumas (Morselli 1984), he said, equal volumes of all bodies in the gaseous state at the same temperature and pressure contained the same number of atoms; but Berzelius and most continental chemists disagreed with this and accepted only that equal volumes of the permanent gases contained equal numbers of atoms. The latter was a simpler rule and it avoided the anomalies which had made Dumas despair of the atomic theory (Buchdahl 1959). But the British school of chemists, Johnston noted, had adopted a third relationship which was identical with that of Berzelius except that oxygen was made to contain twice as many atoms as an equal volume of any other permanent gas. The most obvious consequence of this rule was to give water the formula HO instead of Berzelius's H_2O. 'This opinion', which was due to Thomson, 'involves a departure from the supposed simplicity of nature, for which there is *a priori* no sufficient reason', declared Johnston. Such a rule had the singular merit that it produced simpler formulae for oxygen compounds but, Johnston warned, a good deal of further research was needed to see whether it was the true relationship or only a convention. As we have seen, although Prout had adopted Thomson's system in his anonymous papers, by 1832 he had adopted a molecular system similar to that of Avogadro and Ampère.

Whatever the relationship was between specific gravity and atomic weight, Johnston recognised that specific gravity measurements were as much a bone of contention among chemists as atomic weights. To correct the latter by means of the former would therefore probably only lead to a magnification of errors. Clearly separate experimental programmes would have to be undertaken in order to determine accurate values for both quantities, and their mutual relationship would follow from this. It was for this reason, explained Johnston, that the Chemistry Committee of the British Association had made its two recommendations. Finally, Johnston advised that the case for or against the Prout–Thomson law of multiple weights remained unproven.

In 1833 Johnston could, perhaps, have been more dogmatic. For in his Oxford paper Turner had stated only that the atomic weights which he presented were 'nearly correct', and that the integral weights hypothesis was 'inconsistent with the present state of chemical knowledge'. These points were elaborated in a paper read to the Royal Society on 16 May 1833. In some ways, however, Turner (1833) appeared less certain of his conclusions in this paper. For example, he went back on some of his quite accurate earlier results; and though he offered specific results, his individual determinations were often discrepant within the range demanded by Prout's law.

	Thomson (1825)	Turner (1832)	Turner (1833)	Berzelius
Pb	104	103.5	103.6	103.5598
Ag	110	108.0	108.0	108.1285
Ba	70	68.7	68.7	68.5504
Cl	36	35.5	35.42	35.412
N	14	14.0	14.15	14.1628
Hg			202.6	202.5315
S			16.09	16.0932

His values for silver ranged from 107.92 to 108.08, which led to the slightly embarrassing average integer, 108. He warned that his value for nitrogen was in doubt because any error in the determinations of silver, lead or barium would have been multiplied five-fold in his estimation for nitrogen. To overcome this chemists would have to 'give preference to the more direct method, founded on an exact determination of the densities of oxygen and nitrogen gases, such as we anticipate from the labours of Dr. Prout'. This seems to be a clear indication that Prout continued with his work on specific gravities.

Despite various discrepancies, Turner's values were closer to those of Berzelius than to Thomson's; in these circumstances Turner declared that:

> Dr. Prout's hypothesis, as advocated by Dr. Thomson,—that all atomic weights are simple multiples of that of hydrogen,—can no longer be maintained.

Integral numbers might still be used as convenient approximations for 'medical men, students, and manufacturers', but otherwise chemists must not delude themselves that it was a scientific truth. However, Turner (1833: 544) wanted to salvage something.

> Let me not however be misunderstood: I mean simply to affirm that the experiments by which it has been attempted to prove the truth of this hypothesis are inaccurate. I may go further, and declare it to be not

only unsupported by evidence, but to be at variance with the most exact analytical researches which have been conducted. I deny not that some simple relation subsists among atomic weights, and that their ratios may possibly be expressed by some simple series of numbers; but at present no one has assigned any physical cause for the existence of such a relation; no such relation has hitherto been discovered; nor, as appears to me, has analytic chemistry attained that degree of perfection which can justify any one in finally asserting or denying its existence.

Turner has sometimes been criticised for this rather wavering conclusion; but coming as it did from a man who took a strictly positivistic line towards the atomic theory, this empiricism was only to be expected. Moreover, Turner was surely right, for his point was that since analytical chemistry was open to further improvements, it would be wrong of chemists to make unqualified positive or negative guesses about mathematical relationships between the elements. Yet it is doubtful if Turner would have been satisfied by a periodic law, for he also declared that he required a 'physical cause'—that is, a theory of the elements that would account for any such quantitative relationships. But this only came with the electronic theory of the atom. Turner evidently did not feel that Prout's protyle suggestion was anything more than a speculation, and a speculation that an empiricist should avoid.

It is clear from Turner's remarks that he did not think that his work proved that there was *no* relationship between the atomic weights of the elements. All he had done was to reject the form of Prout's hypothesis upheld by Thomson. Hence it was possible for chemists to continue their support for various forms of Prout's multiple weights hypothesis, and of course its corollary, the protyle hypothesis. No chemist could avoid noticing, for example, that Turner's and Berzelius's values for the atomic weights of the elements on the hydrogen scale were still curiously close to whole numbers, or simple fractions like $\frac{1}{2}$ or $\frac{1}{4}$. Although Thomson refrained from commenting upon the multiple weights hypothesis in his *History of Chemistry* published in 1830 (Thomson 1830–1), he seems to have concluded that Turner had merely subscribed to the views of his continental enemies. In 1836, in 'the last cry of a project-maker' (Liebig 1836), he continued to uphold the validity of the hypothesis even though he no longer used Prout's name in this connection (Thomson 1836). This seems to have been Thomson's last comment on the subject; he lived until 1852, and it would be interesting to have known his reaction to the work of Dumas and Stas in 1840.

Two publications in the *Philosophical Transactions* for 1839 illustrate the problem of falsifying Prout's multiple weights hypothesis. In the first, Frederick Penny (1839) made a careful series of analyses using techniques that foreshadowed those used by Stas. He found in favour of Berzelius and Turner.

> The estimates in general use among British chemists are not the strict representatives of chemical truth, founded on experiment, and . . . the favourite hypothesis, of all equivalents being simple multiples of Hydrogen, is no longer tenable.

Thus, even six years after Turner's great paper it is clear that British chemists were still using Thomson's simple values even for accurate work, and that the integral multiple weights hypothesis was still maintained by them. This is also clear from a paper published by Richard Phillips (1839) in the same year. Phillips—friend of Prout and a continued supporter of Thomson—reached the opposite conclusion to Penny; Turner's results were inaccurate and 'no material, and scarcely even any appreciable error can arise from considering the equivalents of hydrogen, oxygen, azote, and chlorine, as 1, 8, 14 and 36 respectively'.

Penny's table (1839).

	Thomson (H = 1)	Turner	Penny (1839)
O	8	8	8
Cl	36	35.42	35.45
N	14	14.15	14.02
K	40	39.15 ⎱ Berzelius	39.08
Na	24	23.3 ⎰	23.05
Ag	110	108.0	107.97

Even if one agreed with Berzelius, Turner and Penny in the rejection of Thomson's values, it was still possible to support a multiple weights hypothesis. In his own publications before 1825, Thomson had always implied that the basis of a law of multiples could be a fractional unit of the weight of hydrogen. This was also Prout's idea, and in 1831 he proposed that this might have some sort of physical basis. In this year the Professor of Chemistry at Oxford, Charles Daubeny, prepared his *Introduction to the Atomic Theory* for publication, and he sent the proof sheets to Prout for his comments. The latter obliged with a long letter (printed by Daubeny as an Appendix) in which he suggested that there was no sufficient reason 'why bodies still lower in the scale than hydrogen (similarly related to one another, however, as well as to hydrogen) may not exist, of which other bodies may be multiples without being actually multiples of the intermediate hydrogen' (Daubeny 1831: 129). This speculation that the protyle might be half, a quarter, or some other fraction of the hydrogen atom, was Prout's first public pronouncement on the protyle hypothesis since 1816. It was to be taken up by Marignac, Maumené, Pelouze and, above all, by Dumas; but it was to be firmly rejected by Stas.

This new suggestion from Prout prompts the questions: how did Prout view the fortunes of his hypotheses during the period we have reviewed? And more generally, what support did the protyle or reductionist hypothesis receive during the period 1816 to 1839? As we have recognised, the latter question was bound up with the status of the chemical elements.

Since Prout has left no recorded comment concerning his attitude towards the controversies over atomic weights, interpretation will necessarily be conjectural. I think that it is almost certain, however, in view of Prout's letter to Daubeny in 1831, that his commitment to the unity of matter was strong enough to withstand any criticism of Thomson's atomic weights. Like most British chemists he probably felt at first that Thomson was right; in 1827 he noted with pleasure that the law of multiples had been adopted by Thomson and most other British chemists. But the fact that he disagreed with Thomson over the analysis of oxalic acid, and the fact that he helped Turner with his analysis of silver chloride, suggests that he was willing to agree with Turner's strictures on Thomson's methods, and also agree with him that the simple law of multiples maintained by Thomson was erroneous. This would have been the point of his new speculation offered to Daubeny. Indeed, in the Gulstonian lectures which Prout (1831a: 262) delivered to the Royal College of Physicians in 1831, he wondered whether:

> the science of chemistry had not been rather retarded by it [Dalton's theory] than advanced; for to suit the imaginary standards of this bed of Procrustes, real results, I fear, have been too often extended or compressed beyond all legitimate bounds, and thus truth sacrificed to error.

This could refer to Thomson. It is clear from the *Bridgewater Treatise*, too, that Prout never abandoned his own beliefs regarding the unity of matter, or the existence of a mathematical relationship between the elements.

As far as his contemporaries are concerned, there was surprisingly little comment on the question of the status of the chemical elements between 1816 and the establishment of the phenomena of isomorphism and isomerism, and the development of the organic radical theory in the 1830s. Ludwig Meinecke, Professor of Technology at Halle

University, who translated several of Prout's papers into German for *Schweigger's Journal*, published a series of papers on specific gravities and stoichiometry between 1816 and 1819 (Prout 1932: 7–13). He may have been influenced by Prout's first paper of 1815, though he later denied this and suggested plausibly that they had both independently derived their ideas from the work of Dalton and Gay-Lussac. Meinecke adopted a round number (*runder Zahl*) to represent the weight of a portion (*Antheile*) of each element; this number was deduced from the analyses of other chemists but so chosen as to be a multiple of either the portion of oxygen or of hydrogen. He attributed to Dalton the belief that all stoichiometric magnitudes were integral multiples of hydrogen, but like Prout he pointed out that the multiple relations suggested that Daltonian atomism needed modification: there were not *many* elements.

> Even if one does not accept the atomistic views which are the basis for Dalton's rule, one must nevertheless think it curious that most of the stoichiometric numbers are exactly divisible by the number for hydrogen, and all these numbers are nearly so divisible. Although it must not be concluded from this that all substances contain hydrogen as a fundamental element, it has to be assumed that every substance contains a specific proportion of the principle that particularly characterises hydrogen (namely, combustibility, affinity for oxygen, phlogiston, negative electricity) and since every new experience in chemistry shows with increasing clarity that there is a great simplicity in chemical compounds, one may venture the hypothesis that the specific degrees of combustibility (oxidiseability, negative electricity), which characterise the simple substances, can be expressed by masses of the hydrogen value; or, in other words, that since the stoichiometric numbers are actually only derived from their behavior to oxygen, all stoichiometric numbers of the simple substances have to be divisible by the value for hydrogen [Meinecke 1816: 162; Farber 1964].

Consequently, there had to be a simple relation between specific gravities of gases and their stoichiometric values. Meinecke, then, thought that the measurable properties of the elements and compounds suggested either that the 'elements' were graded polymers of hydrogen or that the elements were mixtures of hydrogen and some other principles.

Despite the large number of papers he wrote on the subject, together with priority disputes, Meinecke's work attracted little attention, and his suicide in 1823 prevented him from entering into the later Thomson–Turner discussions. Although much attention was paid to *Naturphilosophie* in Germany, his ideas do not appear to have sparked off any interest in the possible light which chemistry could shed on the unity of nature (Löw 1980). Instead that interest seems to have been the preoccupation of British chemists.

In his rejected Royal Society paper on the kinetic theory of gases, John Herapath (1821) pictured heat as the motion of elastic atoms of various shapes and sizes. Clearly this was a corpuscular theory of matter; not unnaturally therefore, he supported 'one of the sublimest ideas of the ancients'.

> There is but one kind of matter, from the different sizes, figures, and arrangements of whose primitive parts, arises all the beautiful variety of colour, hardness and softness, solidity and fluidity, opacity and transparency &c which is observed in the production of Nature.

Herapath believed, on the grounds of simplicity alone, that probability was 'strongly in favour of the ancient idea'. However, chemists paid little attention to these physical fantasies of Herapath even though he went on to develop a molecular theory (similar to the forgotten Avogadro's) which encompassed chemical reactions as well as the phenomena more familiarly described as kinetic. In this analysis he allowed the particles of the so-called simple bodies to decompose during chemical change (Herapath 1822).

Faraday, who was in many ways a faithful disciple of Davy, expressed dissatisfaction with the great number of metals in 1818. He also, like Davy, adopted a force-atom which supposed a homogeneity, or unity of matter (Faraday 1870: i, 256; Williams 1965). The alteration of force fields explained not only chemical change, but raised the possibility of the transmutation of elements. Later in the century, Faraday became very interested in Dumas's speculations about the elements, but it is not known whether he had any intercourse, social or scientific, with William Prout.

Apart from these two well known scientific personalities, and odd references by one or two other more obscure figures, the interest of chemists seems to have been entirely absorbed by the problems of atomic weight determinations, analysis and the Prout–Thomson multiple weights hypothesis. The question of the simplicity or complexity of the chemical elements was always in the background however. With the appearance of fresh arguments based upon new analogies drawn from the organic radical theory, isomorphism, isomerism, atomic heats and, above all, the redetermination of the atomic weight of carbon by Dumas and Stas in 1840, Prout's protyle hypothesis and the status of the chemical elements once again became a talking point among chemists (*Brit. Ass. Rep.* 1837: 210; Knight 1967).

At the beginning of 1840 bad health sent Dumas to take the waters at Aix where he met the young Belgian chemist, Jean Servais Stas.

Together they began to redetermine the atomic weight of carbon which had proved necessary because of the anomalies in the analyses of naphthalene when Berzelius's value C = 76.42 (O = 100) was used. The experiments of Dumas and Stas were continued in Paris with the greatest care by burning pure carbon compounds in oxygen. There was no doubt about the result. Berzelius's old value was too great by at least two per cent, and it would have to be reduced to C = 75.

In a letter to the French Academy of Sciences Dumas and Stas (1840) pointed out that this value had been the theoretical one deduced by Prout in 1815 (C = 6, H = 1). Had Prout been right in other cases?

> If, as Dr. Prout thought, and as now seems to be quite probable, all atomic weights are multiples integral of that of hydrogen, it would be necessary to rectify many atomic weights.

Indeed, all atomic weights should be redetermined immediately, both for the sake of inorganic and organic analyses, and because of the light the new results might shed on the true nature of substances which had been thought simple.

While not all chemists could share the enthusiasm of Dumas and Stas for the hint of a new chemical and world viewpoint, their experimental result was rapidly vindicated. Already, in 1839, the English chemist George Fownes (1839), working in Liebig's laboratory, had reduced the value of carbon's equivalent, and the need to make some reduction was confirmed by Liebig and Redtenbacher (1841). In the same year the French result was completely vindicated by the Germans Erdmann and Marchand (1841), and by the Swiss analyst Marignac (Partington 1964: iv, 230) who showed that Liebig and Redtenbacher's method was inaccurate. But because of their personality disputes with Dumas over the theory of organic chemistry, neither Liebig nor Berzelius was able to adopt the French result with its implications of what Liebig called *Multiplenfieber* (Liebig and Redtenbacher 1841). Yet Berzelius was sensible enough to realise that his own value for carbon was at fault. In a new determination made by physical methods he found C = 75.12 (Berzelius 1844).

Just as Prout's original papers, or rather those of Thomson, had split chemists into two camps, so did this redetermination of the atomic weight of carbon. Throughout the 1840s, therefore, there was a bustle of analytical activity in which improved methods of accuracy, such as buoyancy corrections and the determination of excess solutions after volumetric endpoints, were devised as Dumas, Stas, Erdmann and Marchand, Pelouze, Maumené and Marignac buckled down to the redetermination of atomic weights. Oddly, this was almost exclusively a continental activity, and British chemists, though intensely interested in the results, preferred to speculate rather than investigate.

Erdmann and Marchand, who made their determinations between 1841 and 1852, were in favour of the integral multiples hypothesis, and the latter used integral weights in his biochemistry textbook, *Lehrbuch der Physiologischen Chemie*, in 1844. Dumas (1842) himself immediately went on to redetermine the ratio between hydrogen and oxygen, and concluded that $O = 8$ ($H = 1$) within the limits of experimental error. The result was confirmed by Erdmann and Marchand (1842). Dumas (1842) now felt convinced that Prout had glimpsed some great general truth.

> In considering water to be composed of 1 hydrogen for 8 oxygen, a chemist will never be exposed to an error in his experiments or calculations . . . all the atomic weights are in need of attentive revision; without adopting or rejecting the opinions of Doctor Prout, I am forced to allow that they are generally in accordance with my experiments.

Prout's hypotheses were once again open questions.

Both Dumas and Stas began their work with a bias towards the integral multiple weights hypothesis and the unity of matter; but from 1840 onwards their paths diverged. In view of Dumas's commitment to a molecular theory of matter, his feeling for the complexity of the chemical elements and that there must be mathematical relations between them is not surprising (Dumas 1837: 304, 346). In particular he pointed out that specific heats and isomorphism were evidence for the complexity of the elements. 'In my youth', Stas reported to Leo Baekeland,

> I was an ardent believer in the unity of matter as expounded by Prout. I was so well convinced about his theory that I became eager to furnish additional proofs by redetermining more accurately the atomic weights of those elements where the atomic numbers were not an even multiple of hydrogen. I simply imagined that more careful determinations would have eliminated these irregularities. But the more I eliminated any errors of experimentation, so much the more did my results contradict my dearest hopes. Finally, I had to admit that I was beaten and had spent the most important part of my life in killing my first love as a theory [Kendall 1949].

Stas, in Belgium, spent twenty years from 1840 to 1860 working towards this conclusion; but Dumas (1859) in France, spent seventeen years from 1840 to 1856 working towards the opposite conclusion that Prout's law of multiples was true—in a *modified* form. Unfortunately, Prout did not live to read these contradictory views.

During the 1840s there was also renewed speculation in Great Britain concerning the possibility of transmutation. In 1844, David Low, Professor of Agriculture at Edinburgh, published a very successful *Inquiry into the Nature of the Simple Bodies of Chemistry* in which he endeavoured to show that:

> It is not necessary, in our inquiries into the phenomena of chemical actions, and the laws which determine them, to assume the existence of many elements, distinct in their corpuscular constitution, from one another, and from the bodies which experiment has determined to be composed of more than one element or member.

Low felt as unhappy as Davy once had about the apparent complexity of the chemical world, and like him supposed that analogical arguments implied that:

> If bodies which we term simple, present the same general physical properties, and exert the same chemical actions as those which we term compounds, and pass into the compound bodies in their characters and functions, the merely negative evidence, that we are unable to decompose them by overcoming their chemical affinities, should not invalidate the conclusions, that *both* kinds of bodies are to be placed in the same class of natural products, and cannot be separated the one from the other by so wide a chasm as a distinct corpuscular constitution [Low 1848: 2].

This was sober reasoning in the Davy–Prout tradition. But Low, drawing upon the analogy suggested by the radicals of organic chemistry, went on to speculate extravagantly that all the chemists' elements were compounds of just hydrogen and carbon—the elements with the lowest atomic weights. There was no reason, he thought, why even these two elements should not turn out to be composed of a single matter which was polarised as $m+$ and $m-$.

Richard Phillips (1844), who otherwise supported the multiple weights relationship, wrote a very sarcastic review of Low's book; but Low replied reasonably enough that his ideas concerning the complexity of the elements were not new. He had merely 'ventured to enunciate the proposition somewhat more precisely' than, for example, Davy; Low did not mention Prout. Some experimental evidence was cited by Low and by his fellow countrymen Rigg (1844) and Samuel Brown (1846). Such evidence, which is reminiscent of Miers' work on nitrogen earlier, was both inaccurately done and inaccurately interpreted; as such, it was often refuted (Knight 1978).

Even Berzelius (1823: 39) was drawn into the new debate over the nature of the elements and their arithmetical relations. In 1823 he had published a discussion of the integral multiples hypothesis, referring to it for the first time as 'der Hypothese von Prout', and admitted that it might be true for the lighter elements, though he warned that Thomson's experiments published in the *Annals of Philosophy* were not to be trusted.

> The fact that at present there is not a single reason, either chemical or physical, for believing this to be the case, does not preclude its possibility.

Even so, he pointed out, there were the glaring anomalies of carbon and chlorine.

Thomson's work soon put paid to any benevolent feelings which Berzelius might have had towards Prout's hypothesis. His demand for rigorous analytical standards meant that adjustments to atomic weights were to be made only on the basis of experiment, and not on the preconceptions of Prout's integral 'law'. Turner's work had apparently proved Berzelius right; but then came Dumas and Stas's adjustment of the weight of carbon, together with the fresh determinations of Erdmann and Marchand, and Marignac; analysts who were all committed to a multiples relationship and a unitary theory of matter, but who were respected by Berzelius. Berzelius felt bound to speak out. Therefore he welcomed the inquiry as to his opinion of Prout's hypothesis which was made in 1844 by the American, Benjamin Silliman Junior (Berzelius 1912–36: iii, 248).

Berzelius's masterful reply was published in the *American Journal of Science* in 1845. In it he reiterated his belief that experiment, and only experiment, could decide the issue. When Prout had 'imagined' his hypothesis originally, it had seemed that atomic weight determinations could never attain great accuracy, but that they suggested that, within the limits of unavoidable experimental error, atomic weights were multiples of the atomic weight of hydrogen. Even so, argued Berzelius (1845), this extrapolation had not even then been justified in all instances. Prout, he said mistakenly, had not expressed any idea on the cause of the multiples; but this, if it really existed, could only be that hydrogen was the ultimate primary matter. Where was the evidence for this though? Had transmutations of the elements ever been observed? At the most, chemists had found elements with the same atomic weights; and Berzelius felt that this phenomenon was evidence against a polymer hypothesis rather than for it.

Thomson's experimental work, although quite uncritical, had led to the adoption of the integral multiples hypothesis in many quarters, and the idea had persisted despite Turner's work. Dumas had published atomic weights for carbon, oxygen, hydrogen, nitrogen and calcium which supported the hypothesis, and the incomparably exact work of Marignac (1843) had shown that other elements approached multiples of half the atomic weight of hydrogen. However, to say that the elements were exact multiples of 0.5 would be to go beyond the limits of experimental error. Even Marignac had been forced to conclude that chlorine was an exception to any such law (Berzelius 1912–36: iii, 214). Berzelius's own atomic weights, though some had been made over thirty years before, provided innumerable exceptions to Prout's relation. Yet Berzelius was prepared to admit that some kind of arithmetical law might be found to hold between the platinum metals.

Berzelius's American article shows that he believed firmly in the reality of many elements, and that he felt that the fact that many atomic weights approached integral or half-integral values was largely coincidental. In those cases where elements had very similar properties such relationships might well prove the case, though only experiment could prove them. Berzelius's opinions seem very similar to the conclusions that Stas published in 1860 (Prout 1932: 41). They were reiterated in the last French edition of his textbook where he warned that unitary theories of matter were speculative and undemonstrated (Berzelius 1845–50: i, 15). His final words on the subject are contained in the fourth volume of this translation published in 1847. Speculation on the whole subject had to cease, otherwise:

> It could easily lead to speculations whose falsity would not be revealed until much later. I think, therefore, that in no case, when the atomic weight of a simple body approaches being a multiple of another, should the number found by experiment ever be made equal to this multiple. This principle must be adopted until such time when science has reached the point when confident responses could be made to the accuracy of these relationships [Berzelius 1845–50: iv, 512].

Despite their lip-service to Baconism, few British chemists were this empirical. Daniell (1843: 349) noted that 'the more accurate our analyses become, the nearer appears to be the coincidence of facts with [Prout's] theory'.

> The notion is doubtless founded upon a sense of that sublime symmetry and simplicity which, the more we inquire the more we find pervading all the works of creation; and when we recollect that in the most perfect of all the sciences, the Laws of Kepler themselves, which have been so amply confirmed by the triumphant progress of Astronomy, were derived from similar views of the geometric harmony of Nature, we are inclined not to reject this view of numerical harmony in the compositions of bodies, until actually proved to be inconsistent with the positive results of accurate experiment.

There is a refreshing intellectual honesty about the writings of most of those who supported Prout.

By 1850, then, the experimentalists' conclusions from their atomic weight determinations were that Prout's integral multiples hypothesis did not hold in all instances. At the same time, Marignac (1843; 1902: i, 96), Maumené (1846) and, to some extent, Pelouze (1845), attempted to save the phenomenon by following up Prout's earlier suggestion that the prime matter could be something less than the hydrogen atom by proposing that a unit of $\frac{1}{2}$ might solve the persistent anomaly of chlorine. Marignac (1860), whose accuracy as an analyst was exceeded only by Stas, was prepared to suggest an even lower unit if necessary. When Stas dismissed Prout's law as an illusion in 1860, he urged caution and in its defence speculated that the minute subatomic particles of the primordial matter from which chemical elements were supposedly composed might not necessarily obey the law of mass addition of macroscopic matter. Thus, for an element like silver whose atomic weight (according to Marignac) was 107.921, the difference from 108 of 0.079 represented the loss of mass due to the 'packing' together of the submolecules within the silver atom. This extraordinary, and at the time, fantastic, speculation only bore fruit over 50 years later. It is a clear indication, however, that, unlike Stas, Marignac could not accept that chance alone was responsible for the proximity of most atomic weights to integers. This was to prove a statistically sound belief (Strutt 1901).

Thomas Graham, a firm believer in the unity of matter (Graham 1863), was very excited by these suggestions, and he concluded in 1850 that:

> It appears to be definitely settled that the equivalents of the elements are not, without exception, multiples of the equivalent of hydrogen. The number of chlorine (35.5) is conclusive against that hypothesis. At the same time, the accurate determinations of the equivalents of chlorine, silver, and potassium, by Maumené, lend positive support to the opinion that these and all other equivalents are multiples of half the equivalent of hydrogen. So do the recent determinations of carbon and hydrogen in reference to oxygen, and those of nitrogen, sodium, iron and calcium. The number for lead also, upon the determination of which extraordinary pains have been bestowed by Berzelius at different times, namely 103.56, is favourable to the same view [Graham 1850: 111].

Thus despite the opposition of Berzelius, the last years of Prout's life and the following decade saw a speculative and experimental renaissance of Prout's hypotheses. Not surprisingly, in his last word on the subject, Prout (1845: 144) was delighted that his view on the relations between the elements seemed vindicated.

8

Identical Outsides but Different Insides

> If we be curious to know what matter is, we plunge at once into that deep which surrounds us on every side, and which never yet was fathomed by human intellect [J F Daniell 1843].

One of the outstanding theoretical advances in the first quarter of the twentieth century has concerned our understanding of the atom; and, in particular, how slight differences in the internal structure of atoms may confer either different chemical properties (and so produce very different chemical elements) or the same chemical properties while differing only slightly in mass. The latter structures are called isotopes. Although the concept of isotopy had been foreshadowed by Crookes (1886) and Carnelley (1886) in their respective speculations concerning the evolution of the elements, the term 'isotope' was first introduced by Frederick Soddy in 1913 in order to describe a phenomenon associated with the transformations of radioactive elements. Working consciously within the Proutian tradition, Soddy was aware that unless a simplifying assumption was made, the number of radioactive elements was likely to overwhelm and destroy the simplicity of the periodic table.

It has happened repeatedly in the history of chemistry that when the number of elements has been experimentally increased there has been a simultaneous move to decrease their number by the introduction of a simplifying theory of matter. Between Stas's rejection of Prout's hypothesis in 1860 and Norman Lockyer's first announcement of the 'dissociation hypothesis' in 1873 six authenticated new elements were isolated; and by 1886, when William Crookes announced his theory of the genesis of the elements, the number had risen by a further half dozen (Weeks 1968: 868–9). For Crookes, as he explained in superlative visionary addresses to the British Association for the Advancement of Science in 1886, and to the Chemical Society in 1888, the similarity between the rare earth elements was convincing evidence that they, at least, originated from a common matter. The idea of the chemical element, he suggested, had to be expanded to take into account the acceptance of the periodic law of the elements, and the speculations and experimental arguments of innumerable chemists since the time of Prout concerning the probable complexity of these undecompounded bodies.

Crookes's address of 1886 may be regarded as the culminating point of nineteenth-century speculations concerning the elements, their interrelationships and their evolution. He was a great syncretist of other scientists' hints and suggestions, and by weaving them together imaginatively with the aid of his literary adviser, Alice Bird, he acquired a well deserved reputation as a Victorian sage (Holloway 1965). Not that he falsely claimed to have originated all of the ideas that he expressed so ably; indeed, in the case of the evolution of the elements he drew attention to a number of authoritative predecessors, principally the philosopher Herbert Spencer (1891: i, 1–17), the Oxford sceptic of atomism Benjamin Brodie (1867), with whom he might have coupled Thomas Sterry Hunt (1867), and the astronomer Norman Lockyer (Meadows 1972).

Lockyer and Crookes were of a remarkably similar mould. Firstly, both were endowed with a speculative streak which often led them beyond laboratory evidence. Both had a fine ear for catching the ideas of their contemporaries, and a courageous ability for moulding heterodox views into evocative and plausible hypotheses. Although much of the detail of their common vision of the genesis of the elements was palpably incorrect, their evolutionary ideas were still influential in the new world of atomic physics before the First World War. Both men could be labelled 'Victorian sages' in so far as they were endowed with a philosophical or cosmic awareness verging upon the metaphysical; because their scientific and literary activity was prodigious, as was their ability to range over many different sciences; and finally because they loved controversy and were catalysts of others' work. As H E Armstrong (1928) said of Lockyer, he was a 'seer ... irrepressible, overflowing with energy and enthusiasm, [and] at times displaying an overmastering tendency to fill the picture, often making the rashest assertions'.

Secondly, both were observationalists, experimentalists and practical men of consummate skill; nevertheless, they were greatly indebted to the work done for them by devoted but little acknowledged research assistants. Thirdly, both were founders and editors of important scientific weekly journals which struggled for existence until the early 1900s: Crookes founded and edited *Chemical News* from December 1859 until 1906, and was nominal editor from then until his death in 1919; Lockyer founded and edited *Nature* from November 1869 until 1919. Both men were able writers, publicists and scientific popularisers—though they may not have written everything that was published under their names.

William Crookes (1832–1919). As editor of *Chemical News* he published many speculative papers concerning the compound nature of chemical elements.

Lockyer's contribution to Crookes's imaginative synthesis on the genesis of the elements was to offer supraterrestrial support for the complexity of the elements by way of his dissociation hypothesis

(Dingle 1928). Lockyer's astronomical reasons for postulating dissociation underwent appreciable change with time, and since the hypothesis itself underwent considerable changes at his hands because of a better understanding of the spectroscopic evidence upon which it was based, and because of Crookes's arguments for the complexity of the elements from phosphorescent spectra and chemical fractionation, it is convenient to distinguish between Lockyer's first dissociation hypothesis of the period from 1873 until 1887 (expounded in its most convenient form in his *Chemistry of the Sun* (1887)), and a second dissociation hypothesis dating from 1897 onwards found most conveniently in his *Inorganic Evolution* (1900). Although spectroscopists like Kayser (1899) and physicists like J J Thomson (1897) looked favourably upon the second version of the theory—which was effectively a theory of ionised particles—it will be the first dissociation hypothesis with which we are concerned here.

The term dissociation was first used in its present scientific sense by Henri Sainte-Claire Deville in 1857 when he ascribed vapour density anomalies of substances at temperatures of above 1000°C to the presence of mixtures of dissociation products, and not to the original single substance (Deville 1857; 1860). Lockyer (1874a) first noted the possible significance of Deville's work in a report on contemporary notions of the sun's structure which he made in 1865. According to Wurtz (1865) dissociation referred at first to the phenomenon of the spontaneous decomposition of a substance by heat, and only later to the partial and gradual decomposition which many substances undergo when they are exposed to a temperature below that at which they decompose in bulk. Not unnaturally Deville and his school believed, by the principle of sufficient reason, that all stable compounds would be dissociated at sufficiently high temperature. Support for this view came from the early astrophysical work of William Huggins and W A Miller in 1864; their spectroscopic observations revealed no signs of chemical compounds in the sun.

By 1873, at least, Deville was prepared to go further and to ask: 'What guarantee is there that hydrogen is a simple substance, and that at temperatures of millions of degrees it will not split into two elements which can combine to form hydrogen?' (Oesper and LeMay 1950). It is possible that he voiced this conjecture through the stimulus of conversations with the American geochemist, Sterry Hunt, who was a frequent visitor to France and who formed a significant link between the inorganic development theories of natural philosophers like Oken, positivists like Deville and Brodie, the spectroscopic interpretations of F W Clarke in America and Lockyer in England and, of course, Crookes, who published their work (Clarke 1873; Hunt 1887; Farrar 1965).

Just as Francis Aston, using positive ray analysis, was one of the first to question the Daltonian assumption that all the atoms of one non-radioactive element were the same, so Lockyer in the nineteenth century, using spectrum analysis, was the first to argue seriously that elementary spectra were not produced by the complex vibrations of homogeneous atoms as physicists then assumed (Maxwell 1875; McGucken 1969), but by the vibrations of the heterogeneous parts of these atoms, many of which parts were common to several of the elements. For, according to Lockyer (1887) 'the spectrum we have observed in each case is not the record of the vibrations of the *particular* substance which we have put into the arc and with which we have imagined ourselves to be working alone, but of all the *simpler* substances or forms of the same substance produced by the short or long series of the "separations" effected'.

Lockyer's principal argument was methodological and consisted of an appeal to the principle of uniformity of nature and to the associated principle of continuity. 'To the Spectroscopist all Nature is one', he wrote (Lockyer 1886). At its simplest level the argument was as follows. If dissociation of compounds can be made to take place at low temperatures in a Bunsen burner, *ergo*, it must also happen at transcendental temperatures in the sun where it is plausible that 'elements' will also dissociate.

> The question, then, is an appeal to the law of continuity, nothing more and nothing less. Is a temperature higher than any yet applied to act in the same way as each higher temperature which has hitherto been applied has done? Or is there to be some unexpected break in the uniformity of Nature's processes? [Lockyer 1887: 265].

However, as so often happens when the analogical argument is used, Lockyer unwittingly confused analogy with identity. For it did not follow necessarily that the same kind of dissociation occurred in the sun as in the terrestrial laboratory—as a reflection on the difference between chemical decomposition and ionisation shows. But Lockyer clearly expected dissociation of the elements to produce simpler elements, i.e. ones of a lower atomic weight. In the absence of a concept of, and evidence for, the electron, and a lack of training in physics, Lockyer could hardly have reasoned otherwise.

The argument from uniformity, and especially the argument from continuity, had been deep-seated in Victorian scientific thought ever since the debates over geological explanation in the 1830s. These debates, as Faye Cannon (1960; 1960–1) showed, were concerned with wider issues than geology and involved the nature of natural philosophy and its relations with metaphysics as much as the structure of the earth. It was William Whewell (1847) who had insisted that in the search for the causes of historical events, just as for mechanical ones, philosophers were bound by the principle of continuity:

> A quantity cannot pass from one amount to another by any change of conditions, without passing through all immediate degrees of magnitude according to the intermediate conditions.

Arguing within a non-evolutionary context, he had supported the amalgamation of actualistic principles with palaeontological progressionism; however, with the rapid development of biology and geology in the 1840s, continuity became a valuable weapon for the justification of inorganic and organic evolution in the hands of Lockyer's seniors and contemporaries, Baden Powell (Knight 1968a), Robert Chambers (1844), Hunt (1882), Stewart and Tait (1875) and Crookes (1912).

That there was surprisingly little opposition to the use of the principle of uniformity in astrochemistry is undoubtedly due to its methodological familiarity. By 1860, Newton's second rule of reasoning in philosophy had become an unquestioned assumption in scientific discourse. Thus once the variability of spectra was understood, attacks on Kirchhoff's identification of terrestrial elements in the sun are difficult to find (Crookes 1861). Although some astrophysicists did argue against uniformity and imagined 'space divided into chemical parishes' (Lockyer 1900), the only chemists' objections were voiced by Berthelot and Mendeleeff. The latter argued in his *Faraday Lecture* of 1889 that Lockyer's analysis of the iron spectrum did not prove dissociation, but only that the iron experienced the same changes in the sun as in the terrestrial arc. Nor would Lockyer's results, if correct, prove the evolution of the elements, for they would only indicate an increase in their number (Mendeleeff 1905).

In a debate on the complexity of the elements at the Academy of Sciences in Paris in 1873, which was precipitated by Lockyer's communication of the dissociation hypothesis (Dumas 1873), Berthelot (who was not intrinsically opposed to the idea of the unity of matter, as his researches on the history of alchemy show (Berthelot 1885)) raised the objection that since there was a definite relationship between the specific heat of a compound and its constituents, there should also be one between the specific heats of related elements, if they were compound. But this was not the case. Moreover, decomposition of the elements, if it took place, would be a violation of the principle of continuity since 'the decomposition of our simple bodies ... would have to be accompanied by phenomena of a quite different kind than those which bring about the destruction of our compound bodies' (Lockyer 1887). Clearly Mendeleeff's criticism came from his opposition to the unity of matter and Berthelot's because he did not confuse analogy with identity. Neither man denied the uniformity of nature; each used it in a more traditional way than Lockyer.

As a spectroscopist Lockyer had been well educated in the use of the uniformity principle. In 1869, in collaboration with Edward Frankland (who taught him some chemistry) Lockyer exploited the principle in a successful attempt to produce by pressure in the laboratory the distorted spectra of the chromosphere (Lockyer and Frankland 1869). Their success undoubtedly encouraged Lockyer to the wider use of the principle, together with that of the continuity argument. But although he was never to question the mode of argument, he reversed it in an interesting way. Instead of reasoning that *this* is what happens on earth, therefore something similar happens on the sun, Lockyer argued: *this* seems to be what happens on the sun, therefore we must revise our opinions of what happens on earth. It appears that a dissociation of the elements must be occurring in the sun; therefore every time we prepare a spectrum of an element in the laboratory we are probably dissociating the element into its constituent parts.

In 1872, using a technique established by Lockyer, the American astronomer C A Young found that the lines seen most frequently in solar prominences were common to two or more elements. The most probable explanation, he thought, was either that his reference spectra were impure or that resolution was imperfect. A less plausible alternative was that:

> there is some . . . similarity between the molecules of different metals as renders them susceptible of certain synchronous periods of vibrations [Young 1872–3].

Later, in a letter to Lockyer in November 1879, Young announced the resolution of many of the 'similarities' (coincident lines) into doublets. Nevertheless, Lockyer continued to use Young's original statements in such a way as to imply Young's approval of the dissociation hypothesis (Lockyer 1887: 186, 267). In fact, Young would not admit that dissociation—the identity of these synchronously vibrating molecules—was a reasonable possibility (Young 1882: 89), but the hint had been sufficient for Lockyer, with the aid of his research assistant R J Friswell (1908), to elaborate in a *Bakerian Lecture* to the Royal Society the following year. There Lockyer (1874b) announced a 'working hypothesis' which assumed:

> that in the reversing layers [the absorption layers where Fraunhofer lines are produced] of the sun and stars various degrees of 'celestial

dissociation' are at work, which dissociation prevents the coming together of the atoms which, at the temperature of the earth and at all artificial temperatures yet attained here, compose the metals, the metalloids, and compounds.

How much of this hypothesis originated with Lockyer and how much with Friswell is difficult to assess. Between 1870 and 1875 Lockyer served as Secretary to the Duke of Devonshire's important Royal Commission on Scientific Instruction, and he had little time for experimental work which consequently fell entirely upon the shoulders of Friswell. The laboratory notebooks kept by the latter record the rare appearances of Lockyer in their new private laboratory at the School of Science in South Kensington, and show that all of the experimental work and some of the chemical conceptualisation for Lockyer's *Bakerian Lecture* were his responsibility. Later notebooks from 1887 onwards, with other assistants, show that Lockyer assumed a much more active role; even so, as Dingle (1960) demonstrated, Lockyer always assumed the employer's right to call assistants' intellectual innovations his own.

Lockyer appealed to several observations in support of the provisional hypothesis, notably that the physical changes in the shape and position of spectral lines could be correlated with temperature to produce a continuous sequence from the most simple spectrum to the most complex; and by an appeal to the obvious sequence of elements which Huggins and others had identified in various types of star and tentatively related to a pattern of cooling. Both these arguments appealed to the principles of uniformity and continuity; though as Henry Roscoe (1873) cogently objected, since there was no evidence that the temperature of a terrestrial electric spark was not at least as high as that of the sun or the stars, the absence of certain spectral lines from stellar sources was hardly evidence of dissociation. The possibility of masking made it inadvisable 'to argue concerning the absence of elements in the solar or stellar atmospheres because we don't see them'.

Reflection will show that the most impressive argument for dissociation would be to observe some of the spectral lines of one element, A, in those of another, B. We might then argue that B was dissociated into A. This argument can be called the argument from coincidence or blends, since it involves the detailed mapping of spectra in order to search for coincident lines between the elements such as Young had detected. However, a snag arises immediately, for how can we be certain that coincident lines are not caused by impurities. Supposing B is not dissociating into A, but A is really present in the test substance as a minute impurity. Obviously, unless we can eliminate impurities the argument from coincidence is worthless.

Lockyer (1887) solved this problem brilliantly. By placing a convex lens between the spectroscope and the substance undergoing treatment in an arc or spark, Frankland and Lockyer produced unusual spectra with 'long' and 'short' lines. Light from the hot core of the arc or spark, together with light from the cooler envelope which surrounded the centre, formed the middle of the spectral lines' image of the vertical slit of the spectroscope. But the extreme ends of the lines came only from the coolest parts of the arc or spark. Lockyer found that lines which ran the full vertical length of the spectrogram were more persistent than the short lines if the quantity of the sample was decreased, or if impurities were deliberately added. They also showed reversal more frequently than short lines in the solar spectrum, and were clearly formed at lower temperatures than the short lines which only appeared at the extreme temperatures found towards the centre of the arc or spark.

This suggested to Lockyer a way of eliminating impurities from spectra, and incidentally, a fruitful new method of identifying elements in heavenly bodies. (Solar spectra also produced long and short lines when viewed by Lockyer's method. If an element was present in the sun then its long lines were most likely to be present in the solar spectrum.) First photograph a spectrum, say of iron, which we suspect contains lines of various impurities. By an optical device we can also confront the iron lines on the same photographic plate with a spectrum of a likely impurity. Providing the spectra are photographed under the same conditions of temperature, and a maximum dispersion, it can be argued that if the longest lines of a suspected impurity are present on the iron spectrogram it must be impure. In that case, these lines must be eliminated from the catalogue of wavelengths of iron. The laborious process is repeated with all possible impurities until we reach a pure spectrum of iron.

Of course, it was not really the goal of a pure spectrum of an element which interested Lockyer. More interesting, and more significant, he believed, was the fact that the long and short lines were fundamentally different. In many cases no coincident long lines were to be found, and Lockyer pronounced these spectra pure even though they contained large numbers of coincident short lines. It followed, if these short line coincidences were optically irresolvable, that either different elements vibrated at the same wavelengths, or that one element 'contained' or was dissociated into another. If the latter, then it followed that it should be possible to detect fundamental similarities between the spectra of elements which had hitherto been thought quite distinct.

From a modern standpoint, such a procedure is insufficient. Lockyer perhaps successfully removed the fundamental lines of impurities, but many of his coincidences must nevertheless have been caused by ionised impurities, enhanced lines, or just blends from faulty resolution. Indeed, faulty resolution, and the possibility of doublets in particular, led to a prolonged public and bitter private dispute between Lockyer and the Cambridge collaborators G D Liveing and J Dewar (1915). No better illustration is needed of the tremendous changes which were wrought in spectroscopy by the Bohr model of the atom and by the advent of the quantum theory. Only because of the work on spectral series during the 1890s, and his own work with Fowler on enhanced lines, did Lockyer come to realise the flaws in his argument. The result was the second modified form of his dissociation hypothesis in which the enhanced lines were explained in terms of ionisation (Dingle 1928).

Until then, and especially during the years 1874 to 1878, Lockyer and his assistants Meldola, Ord, Starling and Miller laboriously mapped out spectra and eliminated 'impurities'. Friends were prevailed upon to supply the laboratory with ultra-pure samples of metals: W J Russell provided samples of cobalt and nickel which had been prepared by the suicidal Augustus Mathiessen; Roscoe presented samples of vanadium and caesium; Crookes some thallium; Roberts-Austin of the Royal Mint samples of gold, silver and platinum; Hugo Müller some copper; Holtzman some cerium, didymium and lanthanum and George Matthey of the firm of Johnson and Matthey, supplied samples of magnesium and aluminium (Lockyer 1878–9).

Even though Lockyer and his assistants used the largest available dispersion, they recorded a plenitude of coincident lines. In the case of iron, for example, in the wavelength region 3906.00 to 3997.50 alone there were coincidences with lines found in uranium, zirconium, yttrium, molybdenum, tungsten and vanadium. Although Lockyer refused to accept that these were accidental they eventually proved to be due either to impurities or to doublet blending! Nevertheless, even Liveing and Dewar, to whom this demonstration was principally due, commented guardedly in 1881 that it would be odd if some coincident lines did not occur, considering the variety of elements and their modes of vibration. Nevertheless, their unequivocal conclusion was:

> The supposition that the different elements may be resolved into simpler constituents, or into a single one, has long been a favourite speculation with chemists; but however probable this hypothesis may appear *a priori*, it must be acknowledged that the facts derived from the most powerful methods of analytical investigation yet devised give it scant support [Liveing and Dewar 1881].

However, to Lockyer it seemed even odder that of the 62 lines of the pure iron spectrum between the wavelength region 29–40 only 18 lines were non-coincidences. This, he claimed, was evidence of dissociation, for coincident lines were evidently the basic lines produced by common molecular groupings within different atoms. The hypothesis predicted that these basic lines would vary in intensity from one temperature condition to another according to the degree of dissociation; this was, of course, found to be the case, since ionisation was really occurring.

On 12 December 1878, 'at a crowded meeting [of the Royal Society] such as is seldom witnessed' (Armstrong 1878), Lockyer read a paper on 'the working hypothesis that the so-called elements are compound bodies' (Lockyer 1878–9). Unfortunately, no first-, or even second-hand, report of the meeting seems to have survived, though Lockyer implied later that the reception was frosty, even hostile. The popular press, and evidently some of his chemical colleagues, dubbed him an alchemist—a charge he gently rebutted in the next issue of the *Nineteenth Century* (Lockyer 1879c; 1887, viii–ix). According to William Crookes (1878b):

> The general impression among men I have spoken to is that you were badly treated the other night by chemists. I think if you will let me print what you *did* say in next *Chem. News*, it will quite clear up any little doubt there may be in the minds of chemists as to what you did and did not say. Many men think you were dabbling in the Black Art, and were going to bring before them the Powder of Projection. The discussion was on the *Daily News* article of the week before and not on your paper, which a most calm, temperate and philosophical discussion of certain spectroscopic anomalies with a very legitimate suggestion for a solution of the difficulty.

One of Lockyer's more exciting lines of evidence concerned the spectrum of indium. His assistant Raphael Meldola (Eyre and Rodd 1947), who had replaced him on the 1875 eclipse expedition, had found no sign of the familiar h line [Hδ] of the solar hydrogen spectrum in solar prominences. Now Lockyer (1879–80) found this line in the indium spectrum. He concluded:

> This unmistakeable indication of the presence of hydrogen, or rather of that form of hydrogen which gives us the red and blue [hydrogen] lines, showed us that in the photograph of [the indium spectrum] we were not dealing with a physical coincidence, but that in the arc this *special form of hydrogen had really been present*; that it had come from the indium, and that it had registered itself on the photographic plate, although ordinary hydrogen persistently refuses to do so.

This was not occluded hydrogen, as Lockyer convincingly demonstrated; in fact indium possess a line, 4101.775 adjacent to the Hδ line 4101.733.

Henry Armstrong, like Crookes, a believer in the complexity of the elements, provided a sober account of Lockyer's thesis for *The Times* (Armstrong 1878; 1928), but he tempered any private enthusiasm he had by saying that Lockyer's arguments were so unlike those familiar to chemists that they could not possibly be blamed for feeling that he had not proved his case. Chemists, he suggested, would only be convinced of the complexity, or the decomposition, of the elements in one of two ways. Firstly, if physicists guaranteed that the spectral phenomena could not be explained as simply in any other way than that the elements were compound bodies—clearly before the advent of the electron this was not possible. Secondly, if Lockyer were to provide unequivocal proof of decomposition in the laboratory. Here the proof implied was one which did not rely on the use of spectroscopic evidence, but one which based its case on a weight change.

Stokes (1879), who became sympathetically disposed towards the idea that elements were complex bodies as a result of Crookes's work on electric discharges through gases, also told Lockyer privately that if dissociation of the elements actually occurred in the electric arc then 'we ought to expect to find after continued action of the arc the products of decomposition in sufficient quantity to enable us to detect the presence chemically, at least in the case of elements for which there are delicate means of qualitative detection'. However, he warned that spectrum analysis was far more delicate an analytical tool than the routine of qualitative analysis.

> The only way to answer satisfactorily as it seems to me the objection arising from the supposition of impurities, if you depend on chemical analysis, would be to purposely introduce a quantity of the suspected impurity not too small for chemical detection, and then show, supposing, which is not likely, that the fact is so, that the introduction would not account for the phenomena observed.

Coincidences due to impurities (which almost certainly existed in nearly all of Lockyer's experiments) would have shown up by this method, but not coincidences due to blending. Whether or not Lockyer ever performed Stokes's sceptically conceived experiment, we can see the strength of the dissociation hypothesis in that the results of this examination would have been ambiguous, or even in its favour.

Unfortunately for his reputation, Lockyer was the sort of man to accept the challenge of producing tangible proof of dissociation. Indeed, he had already gone a long way towards forestalling Armstrong and Stokes by setting his chemical assistants the task of trying to distil atoms as if they were a mixture of organic substances. This programme is first mentioned in a laboratory notebook for 22 January 1879, which is about the same time that Crookes and his assistant J H Gardiner began a laborious series of fractionations of the rare earth elements in a futile attempt to find common spectra (Crookes 1884; Dekosky 1973). On the basis of the continuity argument Lockyer had already explained why the hottest stars contained only hydrogen, and this had suggested that it might be possible to dissociate certain elements on earth into hydrogen by using the drastic heating conditions of an electric spark. Accordingly, in June 1879, following hectic activity with Geissler tubes and Sprengel pumps, Lockyer was able to announce proudly that he had succeeded in detecting hydrogen lines in the spectrum of sodium which was undergoing low pressure distillation (Lockyer 1879a). In a letter of 31 May 1879 Crookes congratulated Lockyer on his 'wonderful sodium discovery'; but a week later he reported that he had not succeeded in repeating the experiment.

This case was, in fact, little different from the indium experiment of the previous year. However, at the meeting of the British Association at Sheffield in 1879 Lockyer went much further and stated that he had actually decomposed several elements into hydrogen in measurable amounts (Lockyer 1879b)! Had Prout been right after all? Certainly the results seemed to vindicate Lockyer's conception of what occurred in hot stars. Sodium, he claimed, produced about 20 volumes of hydrogen; phosphorus, magnesium, sulphur and indium produced 70 volumes of hydrogen; while lithium broke the record with 100 volumes. The results may appear absurd (and may be explained away in terms of occluded hydrogen and water vapour impurities), but they help to explain the intense excitement in the same year when it was rumoured that the German chemist Victor Meyer had succeeded in decomposing chlorine (Armstrong 1879; Brodie 1879a; Anon 1880).

Needless to say, in later years Lockyer preferred to forget this episode (Lockyer 1887: 302). No doubt he took to heart the advice of Brodie (1879b), who wrote to him recalling Humphry Davy's similar blunders seventy years previously:

> Be on your guard against chemical impurities. Our methods of separation are only relative to the object we have in view. What is considered pure for one object is impure for another and to procure a metal *spectroscopically* homogeneous must indeed be difficult if not an insoluble problem.

And he went on to hint that a lesson might be learned from Dumas's recent discovery that some of Stas's meticulous atomic weight determinations could be invalidated by the occlusion of oxygen in pure silver (Dumas 1878).

Similar practical advice came from Lockyer's constant friend Henry Roscoe, who confided to his cousin, the economist and former chemist, Stanley Jevons, his agitation over the thought that Lockyer 'had blundered and would have to draw in his horns' (Jevons 1886). Earlier, in 1874, he had warned Lockyer not to compromise his position in the Civil Service by publishing doubtful scientific views. 'If any Government appointment were to be in the field might not some of your friends (?) make capital out of your views detrimental to your prospect?' (Roscoe 1874). To this, Roscoe's Manchester colleague, Balfour Stewart, added the postscript 'Put the papers in some archives to be opened at the general resurrection of the just'.

Roscoe was a most persistent, helpful and sympathetic critic. 'Are *all* the elements knocked on the head yet', he joked (Roscoe 1879). Working at Manchester, or occasionally with Lockyer's assistants at South Kensington, he showed that the thallium samples which Crookes had submitted to Lockyer contained lead, that Lockyer's 'purest' zinc also contained lead; and that the outbursts of hydrogen from sodium and potassium were due either to occluded hydrogen, or to hydrogen from hydroxide impurities (Roscoe 1878a; 1878b; 1878c; 1880). Needless to say, an embarrassed Crookes was able to retrieve his own reputation by tracing the thallium sample and showing that it had not been part of the ultra-pure samples he had used in its atomic weight determination (Crookes 1878b; D'Albe 1923: 295–6).

However, despite these criticisms, and despite the exacting experiments of Liveing and Dewar, Lockyer went on 'chasing elements', largely because the dissociation hypothesis still fitted the non-chemical and purely astrophysical evidence extremely well. For example, the height levels of the elements in the sun bore no obvious relation to their atomic weights unless the sun was regarded as a multilayered dissociation furnace; the spectra of sunspots and of prominences could be (erroneously) correlated with temperature, and hence with dissociation; and, in Lockyer's view, the flash lines seen in the solar eclipse of 1882 correlated with predictions made by the hypothesis (Lockyer 1887: 271–2). Finally, the demonstration by Abney and Festing (1881) that organic radicals could be detected by their infrared absorption spectra suggested that his own attempt to recognise inorganic 'radicals' within elements was not inherently absurd. The extremely comprehensive way in which Lockyer related the dissociation hypothesis to a variety of astronomical problems served only to increase the number of possible areas of criticism; and astronomers and physicists like Huggins, Young and Schuster raised powerful objections from within their own fields (Dingle 1928:

227–53, 292–315; Meadows 1972). Together the formidable united attacks from astronomers, chemists and physicists made Lockyer into an iconoclast—a position made bearable only by the respect shown to him by the scientific community as editor of *Nature*.

Nevertheless, convinced that his dissociation hypothesis was right, in 1887 he published *Chemistry of the Sun*, which was reviewed enthusiastically by Crookes (1887). Although he carefully played down the evidence from the decomposition experiments, he utilised all the familiar analogical arguments which chemists had used in the past to suggest the complexity of the elements (Knight 1967; 1978), and he developed a suggestive model of their evolution and of the cause of their spectra. Dumas, Lockyer's chemical confidant at the French Academy of Sciences, had made much of the analogy between hydrocarbon series and the relations between the equivalent weights in the days before the enumeration of the periodic law. Now, and following the speculations of Thomas Carnelley (1886; Knight 1970) in *Chemical News*, Lockyer exploited hydrocarbons as a model of building up an imaginary, but plausible, series of molecular groups whose individual vibrations might conceivably produce the observed confusion of spectral lines.

Suppose there exists a hot star which contains two basic substances, a and b. Then, as the star cools, a and b may associate to produce a continuous series of more complicated molecular groups whose properties graduate just like a real series of hydrocarbons which pass from gases to liquids to solids as the molecular weights increase (Lockyer 1887: 260–72).

$$
\begin{array}{ll}
a + b & C + H \\
a + bb & C + HH \\
a + (bb)(bb) = ab_4 & C + (HH)(HH) = CH_4 \\
\quad + ab_2 = a_2b_6 & \quad + CH_2 = C_2H_6 \\
\quad + ab_2 = a_3b_8 & \quad + CH_2 = C_3H_8 \\
\end{array}
$$

(cooling ↓)

He was confident that such an evolutionary theory would explain not only spectra but also variable valency, anomalous vapour densities, allotropy and polymerisation. His final devastating answer to chemical critics was this; if chemists were so convinced of his ineptitude in purifying substances then they were honour-bound to stop quoting so many decimal places in their own atomic weight determinations. And, of course, if they did so, Prout's hypothesis would stare them in the face!

The case of Lockyer presents the historian with an apparent paradox. On the one hand we have a strong and persistent nineteenth-century belief in the complexity of the chemical elements;

and yet, on the other hand, when Lockyer offered chemists what he claimed was fairly unequivocal proof, he was faced by hostility. Clearly, chemists felt incompetent to handle Lockyer's astronomical arguments, and as far as the ideas of dissociation and chemical evolution were concerned most of them were not more critical of Lockyer than they were of Carnelley or Crookes. But where Lockyer's experiments were concerned, they felt intuitively that he had blundered, that he did not appreciate the subtleties of chemical purification procedures, and that he was 'rash' and 'slapdash' (Armstrong 1928).

Ironically, both Lockyer and his chemical opponents were right. Yet, unfortunately for Lockyer, by 1897—when he began to reformulate the dissociation hypothesis in terms of enhanced lines—he seemed to have overstated his case so often that chemists either ignored him for the new excitements of radioactivity and x-rays or wearily dismissed him as a chemically-incompetent astronomer whose blunders shaded into the dubious and nefarious activities of the American alchemist and transmutationist Stephen Emmens (Kauffman 1983c).

On the other hand, when J J Thomson (1897) published his definitive demonstration that cathode rays were streams of negative particles whose mass–charge ratio was independent of the particular gas in the discharge tube and the order of whose value, 10^{-7}, was 'very small compared with the value 10^{-4} ... the value for the hydrogen ion in electrolysis', he was immediately prepared to cite both Prout and Lockyer in support.

> The explanation which seems to me to account in the most simple and straightforward manner for the facts is founded on a view of the constitution of the chemical elements which has been favourably entertained by many chemists: this view is that atoms of the different chemical elements are different aggregations of atoms of the same kind. In the form in which this hypothesis was enunciated by Prout, the atoms of the different elements were hydrogen atoms; in this precise form the hypothesis is not tenable, but if we substitute for hydrogen some unknown primordial substance X, there is nothing known which is inconsistent with this hypothesis, which is one that has been recently supported by Sir Norman Lockyer for reasons derived from the study of stellar spectra [Thomson 1897: 311].

Thus, by the end of the nineteenth century the experimental work of physicists seemed to be vindicating chemical and astronomical speculators and spectroscopists who had been inspired by the Proutian dream of finding a more fundamental kind of matter. In both the sun and in the intense electric field of a discharge tube gases were being dissociated and broken down, not into ordinary chemical atoms (as Lockyer had at first believed), but into primordial atoms, protyle, corpuscles, a fourth state of matter, or electrons—the term varied, though that of electron was to be the one that stuck (Stoney 1894).

What, then, was an element? In opening his striking *Presidential Address* to the chemical section of the British Association in 1886, William Crookes had already suggested that this question was the first riddle of chemistry (Crookes 1886: 559). For Crookes, the work of Lockyer, the speculations of Brodie (Brock 1967), the analogies between elements and the hydrogen radicals explored by Carnelley (1886) and on polymerisation by E J Mills (1873) as well as the evolutionary writings of Herbert Spencer (1891), left him in no doubt that they were compound bodies. Moreover, the periodic law implied that their individual similarities and differences of property could not be accidental, while the theory of biological evolution suggested that they too must have developed or evolved along certain fixed lines.

> The array of the elements cannot fail to remind us of the general aspect of the organic world. In both cases we see certain groups well filled up, even crowded, with forms having among themselves but little specific differences. On the other hand, in both, other forms stand widely isolated. Both display species that are common and species that are rare; both have groups widely distributed—it might be said cosmopolitan—and other groups of very restricted occurrence. Among animals I may mention as instances the Monotremata of Australia and New Guinea, and among the elements the metals of the so-called rare earths [Crookes 1886: 561; Knight 1968b: 337].

Not surprisingly, therefore, Crookes went on to rekindle Prout's hypothesis by suggesting that elements were complex bodies that had developed by an inorganic process of natural selection.

Before evolution had commenced, Crookes speculated, there had existed a primary matter which Prout, following the Greeks, had called 'protyle'. (A better transcription would have been 'protohyle'.) Sixty years of atomic weight determinations had firmly ruled out the possibility of identifying this primordial matter with hydrogen. However, Crookes was prepared to suggest that it could be helium, the element which Lockyer and Frankland (1869) had identified spectroscopically and still, then, only known to exist in the sun (Kragh 1982). Many years later, Crookes was to follow J J Thomson and to identify protyle with the electron (Crookes 1902). Then, with remarkable vision, Crookes showed how the elements, with their periodic properties, could be conceived as arising under the action of an imaginary cosmic pendulum swung by the efforts of the forces of electricity and heat (loss of temperature over time), and dampened according to the laws $x = a \sin(mt)$ (an oscillation) and $y = bt$ (a simple uniform motion at right angles to the oscillation), where a, b and m are constants, x is the electric force, y is the temperature and t is the time.

Such equations effectively reproduced the spiral form of the periodic table which Crookes's Irish friend, J Emerson Reynolds (1886), had developed for teaching purposes and published in *Chemical News*. Indeed, since genesis occurred in real space, by adding a third oscillatory motion, $z = c \sin(2mt)$, an oscillation twice as rapid as the first and at right angles to the two former motions, an actual three-dimensional figure of eight model of the genesis of the elements could be constructed such that similar elements were located in a vertical plane on identical parts of the helix (Crookes 1888; Greenaway 1962).

Unlike some contemporary British physicists (Burchfield 1975), Crookes held no brief for a pessimistic scenario in which the universe collapsed in a future 'heat-death'; instead he believed the universe to be in continual creation. The radiant heat from ponderable elementary materials in the centre of the universe flowed towards the periphery where it was transformed into protyle and genesis recommenced. Definite quantities of electricity were given to each element at its genesis, and this electricity determined the element's valence and hence its chemical properties. Atomic weight, on the other hand, was only an accidental measure of the cooling conditions that had prevailed at the moment of the element's conception and it was not, as Mendeleeff had supposed, a determinant of its properties. For, if the cooling conditions had sometimes been irregular:

> elements would originate even more closely related than are nickel and cobalt, and thus we should have formed the nearly allied elements of the cerium, yttrium, and similar groups; in fact the minerals of the class of samarskite and gadolonite may be regarded as the cosmical lumber-room where the elements in a state of arrested development—the unconnected missing-links of inorganic chemistry—are finally aggregated [Crookes 1886: 569; Knight 1968b: 345].

In this way it could be seen that the atomic weights determined by terrestrial chemists really represented an average weight of several slightly different atoms.

> When we say the atomic weight of, for instance, calcium is 40, we really express the fact that, while the majority of calcium atoms have an actual weight of 40, there are not a few which are represented by 39 or 41, a less by 38 or 42, and so on.

This explained not only why atomic weights were so persistently and tantalisingly close to integral or half-integral values, but also why there was a 'close mutual similarity, verging almost on identity' in the rare earth elements. He called these closely-related elements 'meta-elements' or 'elementoids' and with some justification claimed, in 1915, that the concept of isotopes was a vindication of this daring speculation. In fact, although Soddy quoted Crookes to give his isotopic concept historical support, Crookes's model was closely bound up with his imaginative notion of the genesis of the elements and his misinterpretation of experiments with the spectra of the rare earth series (Crookes 1888; Dekosky 1973) to be anything more than an inkling of Soddy's isotopes which were a solution to a different problem upon which it had no direct influence and which only emerged after 1900: that of radioactive series.

Radioactivity—the word was coined by Pierre and Marie Curie (1898)—was first discovered by the Frenchman Henri Becquerel (1896) when he was investigating the source of Röntgen's mysterious x-rays, which were discharged from Crookes's cathode-ray tubes, and their possible links with phosphorescent minerals. With so many other mysterious radiations being reported in the literature at the time, radioactivity aroused little international interest until Marie Curie (1898) showed not only that thorium possessed the same 'ray-emitting' property as uranium, but that three further powerful emitters and new elements, radium, polonium and actinium, could be extracted from the uranium ore (pitchblende) (Badash 1979b: 10–12). Although many different explanations for the radioactive phenomenon were produced, it was initially only the ionising power of the radiation which prompted Ernest Rutherford's interest in uranium and its analogues; for, under J J Thomson's tutelage, his first work at the Cavendish Laboratory in Cambridge was on the conductivity of gases in discharge tubes (Thomson and Rutherford 1896). By 1899, however, his interest in the radiation itself aroused, Rutherford (1899) had distinguished two of its components, an ionising portion which scarcely penetrated matter (x-rays) and another which had far less ionising, but far greater penetrating, power (β-rays). The even more intensely penetrating γ-radiation was distinguished about the same time by Paul Villard (1900) in Paris.

Given the model provided by Thomson's determination of the charge–mass ratio of the electron, it was but a small step for physicists to show that β-rays could be deflected by electric and magnetic fields and that their particles were identical to a stream of electrons. The α-rays were more intractable; but by 1902 Rutherford (1903) had shown them to be positively charged particles with a charge–mass ratio $1/4000$ that of an electron. Although he privately speculated that they were helium ions, he was not able to offer any formal experimental proof of this identification until 1908 (Rutherford and Geiger 1908).

But why did uranium, radium, actinium, polonium and thorium spontaneously lose electrons, helium ions and electromagnetic radiation (γ-rays) unlike the vast majority of the elements in the periodic table? And what was the explanation for the bizarre ability of uranium, thorium and radium always to maintain the same level of activity, while polonium and the gaseous emanation (which Rutherford (1900) had found emitted from thorium) usually decreased in activity?

Given 'the growing belief among advanced chemists in the theory that the elementary bodies as known to us are compounds of a unique primary matter (*protyle*) and that transformations of one kind into a similar one is not beyond the bounds of possibility' (Bolton 1898), Rutherford's solution to these questions is not so surprising. His answer, given in 1902 with the British chemist Frederick Soddy, was the theory of the disintegrating atom in which one radioactive element slowly changed into another (Rutherford and Soddy 1902). In this modern form of alchemy, 'thorium, for example, transformed into thorium X, which in turn changed into thorium emanation, which in turn became active thorium deposit', each transitory substance possessing 'its own half-life and its own chemical properties' (Badash 1979b: 16). Although Rutherford and Soddy were privately worried about being accused of reviving the tainted science of alchemy (Trenn 1974), the theory, which fitted the facts so well, received little opposition—a fact which may be attributed to the ground having been so well prepared by Lockyer, Crookes and, of course, the Proutian tradition generally. In any case, support for the theory soon came when, in a classic experiment, William Ramsay and Soddy (Ramsay and Soddy 1903) showed that radium was continuously producing the inert gas helium and when the young German chemist, Otto Hahn, then working in Ramsay's London laboratory, identified radiothorium and its predecessor in the disintegration chain, mesothorium (Hahn 1905; 1907).

Bertram Boltwood, then an assistant Professor of Physics at Yale College, and a regular correspondent with Rutherford (Badash 1969), was working on the thorium disintegration series at the same time as Hahn. Soon after Hahn's report of the discovery of mesothorium had appeared he informed Rutherford (Badash 1979b: 120)

> I have also got a lot of data about mesothorium, which as a matter of fact has *the same chemical properties as radium* [my stress] and can be separated from minerals in the same manner. But I shall hold up this for some time yet in order to give Hahn a chance, for I feel that his priority in the matter should be recognized.

That this recognition of the possibility that the growing number of radio-elements might be reduced, because common chemical properties could be detected, was not more widely noticed is best illustrated by the work of another early American radiochemist. Between 1905 and 1907, H N McCoy, who held a chair of physical chemistry at the University of Chicago, together with a doctoral student, W H Ross, showed convincingly by a long series of complicated chemical reactions that radiothorium could not be separated from thorium (McCoy and Ross 1907). However, although Soddy (1966: 376) many years later saw their work as 'the first definite statement of the complete chemical non-separability of what are now called isotopes', McCoy and Ross pursued their finding no further. As Badash (1979b: 124, 182–5) has commented, 'probably they regarded it as analogous to especially difficult separations among the rare earths, and a way eventually would be found'.

Nevertheless, evidence for the existence of several chemically non-separable radioactive elements soon began to accumulate (Kauffman 1983a). Ionium, separated and named by Boltwood (1907), was found to be inseparable from and identical to thorium; lead and radioactive lead, uranium I and II, and mesothorium and radium formed further examples. In Sweden, using traditional techniques of wet analysis as well as the law of crystalline isomorphism, Strömholm and Svedberg (1909) showed that both thorium X and actinium X were alkaline earth elements and that both were indistinguishable chemically from radium. This led them to suggest that there were three parallel series of radioactive disintegrations occurring in nature which originated in uranium, thorium and actinium respectively. More presciently they went on to speculate that if these series continued through the periodic table then the non-integral character of Prout's law might be explained in truly Crookesian fashion.

> One may suppose that the genetic series proceed down through the periodic table, but that always the three elements of the different genetic series, which *thus together occupy one place* [my stress] in the periodic system, are so alike that they always occur together in nature and also have not been able to be separated appreciably in the laboratory. Perhaps, one can see, as an indication in this direction, the fact that the Mendeleef scheme is only an approximate rule as concerns the atomic weight, but does not possess the exactitude of a natural law; this would not be surprising if the elements of the scheme were mixtures of several homogeneous elements of similar but not completely identical atomic weights [Strömholm and Svedberg 1909; translation by Kauffman 1983a: 4].

It was this paper more than any other which triggered off Soddy's synthesising concept of isotopy. Reviewing the Swedes' work in the

Chemical Society's *Annual Reports* for 1910 Soddy (1911a) opted explicitly for the reduction of the thirty-five or so radio-elements to the twelve available spaces in the periodic table, and hence assumed their identity.

> When it is considered what a powerful means radioactive methods of measurement afford for detecting the least change in the concentration of a pair of active substances, and the completeness and persistence of some of the attempts at separation, which have been made, the conclusion is scarcely to be resisted that we have in these examples no more chemical analogues but *chemical identities* [my stress].

He then moved on, in Crookesian fashion, to the ordinary elements.

> The recognition that elements of different atomic weight may possess identical chemical properties seems destined to have its most important application in the region of inactive elements. ... Chemical homogeneity is no longer a guarantee that any supposed element is not a mixture of several of different atomic weights, or that any atomic weight is not merely a mean number. The constancy of atomic weight, whatever the source of material, is not a complete proof of homogeneity, for, as in the radio-elements, genetic relationships might have resulted in an initial constancy of proportion between the several individuals, which no subsequent natural or artificial chemical process would be able to disturb.

And, still more explicitly, in a paper on mesothorium published in the same year, Soddy (1911b; 1911c) asked rhetorically

> Whether some of the common elements may not, in reality, be mixtures of chemically-separable elements in constant proportions, differing step-wise by whole units in atomic weight. This would certainly account for the lack of regular relationships between the numerical values of the atomic weight.

However, it was not until December 1913 that Soddy (1913a), growing 'tired of writing "elements chemically identical and non-separable by chemical methods"' (Thomson *et al* 1921: 98), coined the word 'isotopes' (Greek 'same place'). Earlier, in March of the same year, the Polish chemist Kasimir Fajans (1913a) had suggested the stellar term 'pleiad' for the same phenomenon, but his term sank without trace, not even being mentioned in Soddy's *Annual Report* for that year (Badash 1979b: 198). On the other hand, Fajans, followed rapidly by Soddy, did enunciate and establish the group displacement laws whereby the disintegrating nucleus of Rutherford and Soddy's radiating atom moved 'into the next but one [family] in the direction of diminishing group number and diminishing atomic mass' (Soddy 1913b: 262) upon emitting an α-particle, and directly into the next family in the opposite direction when a β-ray was emitted (Badash 1979a; 1979b: 194–213). In this fashion the threat posed by the growing number of radio-elements to destroy the elegance and simplicity of the periodic arrangement or to raise the number of 'simple' chemical species to well over a hundred, was masterfully resolved—though, somewhat ironically, Fajans himself, in the tradition of Stas, preferred to regard each and every isotope as a distinct element. However, by the 1920s he, too, had capitulated (Fajans 1923: xii).

Atomic weight determinations by Richards (Kauffman 1983b) of samples of lead from the different radioactive disintegration series, as well as the identification of uranium X_2 with protactinium (element 91) by Fajans (1913b), rapidly offered confirmatory evidence that isotopes were unequivocally and simply varieties of known elements. Moreover, the displacement rules, which effectively showed elements losing two, or gaining one, electric charge as they moved down or up the periodic table, also focused physicists' attention upon the nuclear charge. Between 1913 and 1914, H G J Moseley, a student of Rutherford's who was working at the University of Manchester, showed how the characteristic x-ray spectra of elements could be used to prove that their nuclear charges increased regularly through the periodic table (Heilbron 1974). His suggestion that the nuclear charge, or 'atomic number', was a more accurate reflection of the ordering of the elements in the periodic table than atomic weights was rapidly adopted.

But what experimental evidence was there that non-radioactive elements were isotopic beyond the non-integral values of many of their atomic weights? Did such elements also have 'identical outsides but different insides' (Soddy 1966: 371)? The answer came from that other tradition established by Crookes and furthered by J J Thomson, that of the electric conductivity of gases under very low pressures. It was due to the imagination and resourcefulness of the experimental physicist Francis Aston in building the mass spectrograph which was to lead to the fresh theoretical insights of the 1920s and to the final vindication of Prout's hypothesis as far as the elements themselves were concerned.

Aston was born at Harbonne, near Birmingham, in 1877, the third child of a family of seven children. His father, William Aston, was a metal merchant and farmer and his mother, Fanny Hollis, was the daughter of an important Birmingham gunmaker. Aston was brought up in the comfortable middle-class surroundings of his parents' farm and educated at a local church school until, at the age of twelve, he was sent to public school at Malvern. In the school holidays, however, his parents encouraged him to develop the mechanical and engineering skills which were to be the hallmark of his scientific career by providing him with a workshop and laboratory. In 1893 he entered Mason's College, the forerunner of the University of Birmingham, where he studied chemistry and physics with W A Tilden, P F Frankland and J H Poynting—the latter being a close friend of J J Thomson.

It was by no means predictable at Mason's College that Aston's future work would be in physics. In 1898 he obtained a scholarship which enabled him to work with Frankland on the stereochemistry of dipyromucyltartaric acid esters (Frankland and Aston 1901), while, at the same time, in pursuit of a career, he studied the chemistry of brewing with Adrian J Brown, who had become the first holder of the chair of Brewing and Malting at Mason's College. However, although Aston actually earned his living between 1900 and 1903 as a brewery chemist in Wolverhampton, he returned to the University of Birmingham (formed from Mason's College and other institutions in 1900) in 1903 as a research student with Poynting who, like J J Thomson himself, was engaged in investigating the discharge of electricity through evacuated Crookes tubes. Doubtless research with Frankland on stereochemistry, or routine analytical and quality control in a brewery must have seemed dull by comparison with the opportunities offered by cathode ray research. Not that Aston's chemical training had been wasted. He became a skilled glassblower—a technique which was to prove of the greatest value in the tortuous and vexatious world of high vacuum physics; while, from Frankland he acquired the analytical chemist's feeling for care and accuracy which later became the most obvious feature of his work on isotopes.

What had really caused Aston to abandon chemistry was the publication in 1903 of Thomson's monograph *Conduction of Electricity through Gases*. In this book Thomson had ascribed some of the phenomena observed when electricity was conducted through gases under reduced pressure in a discharge tube as due to the movement of positive ions, i.e. atoms which had been stripped of some of their electrons. The beautiful luminous features in a discharge tube when an electric current is passed are quite distinctive. Around the cathode there is a thin layer of light followed by a variable area of darkness called the Crookes dark space after William Crookes, its first investigator (Crookes 1878a). Beyond this, moving towards the anode, is another area of luminosity (the negative glow) and a second variable dark area, and the first to be described, called the Faraday dark space (Faraday 1838). Finally, there is another luminous area stretching from the latter to the anode itself, called the positive column, which is usually striated with patched of darkness. At still lower pressures the Crookes space enlarges to fill the whole tube which begins itself to fluoresce.

By 1903 these phenomena were well known, but the way in which they varied, if at all, with different gases, at different pressures, with different anode and cathode materials and with different current intensities, was still uncertain when Aston rejoined Poynting at Birmingham. In fact, Aston's first physical research was a detailed investigation of how the length of Crookes dark space varied with pressure and current and in different gases (Aston 1907). Here his prowess at glassblowing served him well, for he succeeded in shaping Geissler tubes with movable aluminium cathodes. In this way the same tube could be made to produce a variety of cathode–anode lengths, enabling him to obtain sufficiently well bounded dark spaces to be able to demonstrate that their length was always inversely proportional to the pressure and the square root of the current. For Aston, this was firm evidence for Thomson's explanation for dark spaces; namely that Crookes space was a region of positively charged particles that were moving towards the cathode. Such a column possessed definite boundaries because the total positive charge within it was in equilibrium with the negative charge on the cathode. When positive ions struck the cathode, electrons were ejected towards the anode with sufficient velocity to ionise any gas particles beyond the region of the equilibrium and so produced the column of 'negative glow'. Alteration of either the pressure or the current disturbed both the number of gas particles available for ionisation and the conditions in the equilibrium area. Either way, the length of the Crookes dark space would vary.

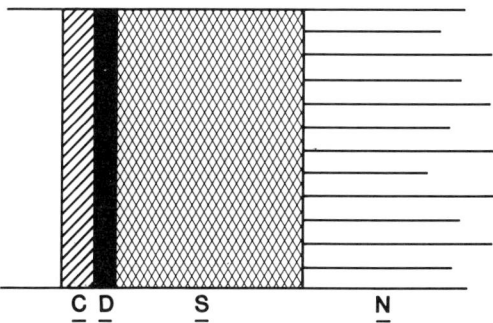

The Aston dark space. A diagrammatic section of the cathode discharge in helium at about 1 mmHg pressure. C is the cathode, D the Aston dark space (c. 0.05 cm), S is the Crookes dark space (c. 1 cm) and N is the negative glow (from *Proc. R. Soc.* A **80** (1908) 46).

It was while working with hydrogen and helium that Aston (1908) discovered a new cathode dark space—an extremely minute layer, scarcely a millimetre in thickness, which bordered the cathode. Although the phenomenon proved very difficult to detect and to measure, Aston contrived ingenious ways of showing that it was independent of pressure (in fact a pressure dependence was subsequently detected) and inversely proportional to the current. Aston's explanation for this phenomenon, which became named the Aston dark space, again rested on Thomson's model of conductivity, namely that it represented the distance fallen by electrons before they had gained sufficient energy to ionise gas particles by collision.

Given the link between Poynting and Thomson, and Aston's use of the latter's work to interpret his experimental findings, it was inevitable that when Thomson required a new personal research assistant in the Cavendish Laboratory he should have asked Aston to fill the vacancy. Although Thomson had investigated the striated positive light of the discharge tube in the early 1890s, the controversy over the real nature of cathode rays and the discovery of the electron forced him to postpone a programme of research on the nature of the positive rays until the early 1900s. Then, with the aid of his assistant, E Everett, Thomson (1907) developed an apparatus whereby a heterogeneous stream of positive rays was filtered through a perforated cathode to be subjected simultaneously to a horizontally placed electric field and a vertically placed magnetic field. By this means Thomson, following the earlier examples of Wilhelm Wien (1898; 1902) was not only able to measure the charge–mass ratio of the particles, but also to render them visible as fairly sharp parabolic trajectories on a fluorescent screen.

According to elementary electromagnetic theory, the displacement due to the electric field, $x = (AeE/mv^2)$ and the displacement due to the magnetic field, $y = (AeH/mv)$ where e is the charge on the particles, v is the velocity of a particle, E and H the respective strengths of the electric and magnetic forces and A is a constant for the apparatus which depends upon the length of the path between the electric and magnetic plates. From these equations a relationship for the resultant parabolic trajectory can be obtained, namely

$$\left(\frac{y}{x}\right)^2 = \left(\frac{Hv}{E}\right)^2 \qquad \text{or} \qquad y^2 = \frac{H^2 v^2 x^2}{E^2}.$$

It follows that although a particular parabolic trace on the screen is made up from particles moving with a variety of velocities, all particles with the same e/m, or the same *mass*, will lie on the same trace. Different masses will trace out different parabolas, and y^2/x is a measure of the e/m.

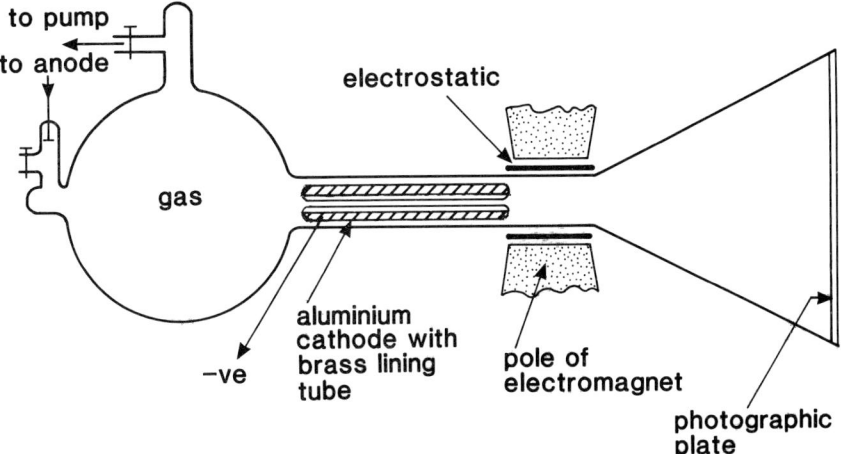

Thomson's positive ray apparatus. A sketch showing the principle of Aston's improved version of the positive ray apparatus (from *Sci. Prog.* **7** (1912)).

Here, then, in Thomson's hands was a powerful new technique for analysing the atoms, ions and molecular species inside the discharge tube (Thomson 1913a), and a device which promised to vindicate once and for all Dalton's assumption that atoms of any one element were identical in mass. At the same time Thomson obviously hoped to resolve the question whether the positive rays (like electrons) were the same for all elements or different, and if the former, whether they could be identified as the protyle. In fact, since positively charged hydrogen always seemed to be present in his apparatus, he was for a time persuaded that hydrogen itself was the fundamental building block of matter (Thomson 1907). However, Wien persuaded him that hydrogen was a ubiquitous impurity (Thomson 1912) and it became one of Aston's principal tasks to improve the design of the apparatus so that traces of hydrogen were eradicated. In this way the more interesting atomic and molecular ions in the tube could be investigated (Aston 1912).

To accomplish this, and to make the instrument into an accurate apparatus for the quantitative determination of mass, Aston replaced Thomson and Everett's first design with a spherical x-ray type discharge tube; he finely engineered the cathode slits; designed a coil for the detection of vacuum leaks from the apparatus; and added a camera with specially sensitive film for photographing the parabolic traces (Aston 1923).

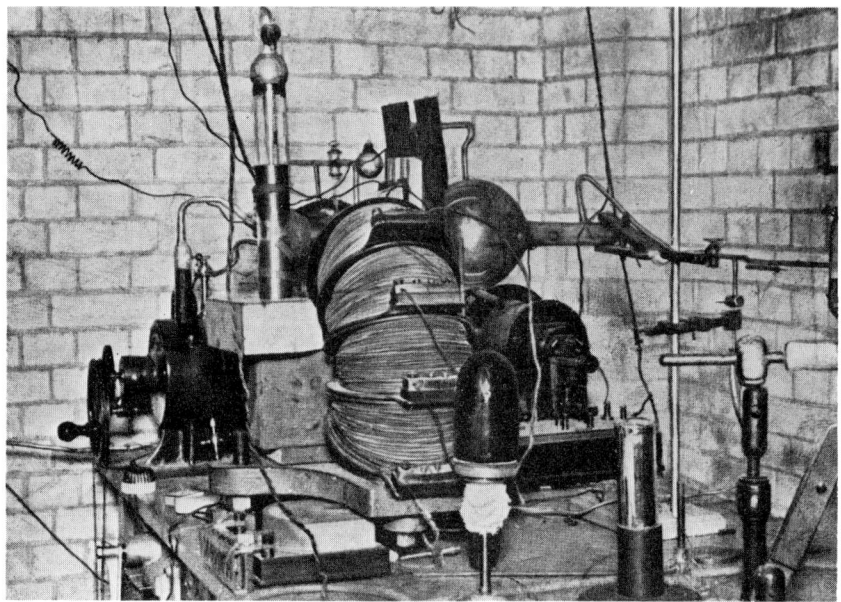

The positive ray apparatus of J J Thompson as improved by Aston (from *Sci. Prog.* **7** (1912) 48).

Probably because Aston's last independent research had involved discharges in rare gases, and because he had samples to hand, one of the first masses analysed by Aston and Thomson in November 1912 was neon. But instead of the single trace that they both expected for this element, neon produced two parabolas, one denoting a mass 20, the other one of 22 (Thomson 1913a). Now, the atomic weight of neon as determined by chemists was 20.2 (O = 16). There were only three obvious and serious possible explanations: either parabola 22 was the ion on neon hydride ($NeH_2^+ = 20 + 2 = 22$) or a doubly charged molecule of mass 44 such as CO_2^{2+}, or it represented an unknown element of atomic weight 22.

Today we know that some of the rare gases can form compounds given the right conditions; but until 1962 no definite compounds had ever been vindicated (Holloway 1968). The zero group in the periodic table was therefore usually referred to as the group of rare or noble gases. Hence, in 1912 it seemed unlikely that neon hydride existed— $(NeH)^+$ has in fact been observed spectroscopically by Pahl and Weimer (1958); nevertheless, Thomson seriously investigated this possibility while leaving Aston to check on the even less likely alternative. Aston took heart, however, from the fact that extremely pure neon, which was unlikely to contain a hydride or foreign molecules, still gave the 22 line; similarly the line persisted when the lines of normal CO and CO_2 were absent. And since Soddy had already declared explicitly that among the many species of radioactive disintegrations atoms of the same element could differ in mass by small whole numbers, it seemed conceivable that a similar phenomenon might occur among those equally extraordinary elements, the inert gases. Soddy (1913b: 266) himself was in absolutely no doubt that Aston and Thomson's work on neon

> appears to be a case of isotopic elements outside the radioactive sequences. ... The discovery is a most dramatic extension of what has been found for the elements at one extreme of the periodic table, to the case of an element at the other extreme, and strengthens the view that the complexity of matter in general is greater than the periodic law reveals. Although the complexity is greater, the problem of atomic structure has been much simplified, because the generalization gives a probable explanation of the absence of exact simple numerical relations among the atomic weights.

Soddy's last comment suggests that he also saw isotopy as an explanation for the non-integral character of Prout's hypothesis.

But to return to Aston. Undeterred by the troubles of radiochemists who had long struggled to effect sensible separations of the disintegration series, Aston reasoned that if isotopes of neon existed it should be possible to separate them. Reflection showed that the capture of the different parabolic beams was scarcely practicable (Lindemann and Aston 1919)—calculations suggested that one might hope to obtain only one tenth of a cubic millimetre of ^{20}Ne and one hundredth of a cubic millimetre of ^{22}Ne per 100 seconds of collection. Fractional distillation seemed a more reasonable proposition and was accordingly attempted with an elaborate glass apparatus of tubes and valves. In this Aston (1924: 39) contrived to separate the supposed lighter atoms of neon by pumping (i.e. evaporating) them into a succession of cooled vessels. Theoretically, the number of lighter atoms pumped off should have been greater than the number of heavier ones in inverse proportion to the square roots of their masses. The process was maintained for about three weeks, or 3000 fractionations, but the result was very disappointing: the atomic weight of the 'lighter' fraction obstinately remained 20.2.

As we now know, of course, Aston failed simply because he did not work this method on a large enough scale. During the Second World War fractional distillation techniques for the separation of uranium isotopes were widely used on an enormous industrial scale.

Since the time of Thomas Graham (1829) it had been known that the velocity of diffusion of a gas was proportional to the square roots of the masses of its molecules. Was it possible, then, that the rates of diffusion of the two different types of atom through porous tubes effect a sensible separation? The method, though workable, was bound to be slow and tedious in view of the very small mass difference between the two types of neon. Nevertheless, although not completely convincing at the time, partial separation was achieved by this method, so that Aston (1913) felt confident enough to announce the existence of a 'meta-neon'—notice the use of Crookes's nomenclature—at the meeting of the British Association for the Advancement of Science in his home city of Birmingham in 1913.

Unfortunately, further research effort was frustrated by the advent of the First World War (Aston 1924: 43). Between 1914 and 1918 Aston returned to chemistry as a technical assistant at the new Royal Aircraft establishment at Farnborough alongside a crowd of future eminent scientists like the physicists F Lindemann and G P Thomson, and the physiologist, E D Adrian. Here he had time to think over the problem of separating isotopes generally and to argue the case with his sceptical messmate, Lindemann, who later became well known to the general public as Lord Cherwell, scientific advisor to Winston Churchill. Although Lindemann played devil's advocate in these discussions, his formidable mathematical ability was put to use in a joint paper on theoretical ways of separating isotopes supposing that they existed at all (Lindemann and Aston 1919).

By the end of the war, when Aston returned to Cambridge, where he was to spend the remainder of his life, he had concluded that a more convincing method of demonstrating the existence of neon isotopes would be to compare the ratio of electric charge to mass of the positive rays of the meta-neon with that expected for neon of atomic weight 20.2. Then, if the less intense 22 line were a neon isotope, the stronger line should be exactly 20, thus giving an average for the mixture of 20.2. He recognised, however, that the method of positive ray analysis which he had helped to develop before the war was lacking in the necessary precision for such a delicate measurement. Rays were lost by collisions in the canal tube, which then silted up or gave rise to unwanted negative parabolas from the capture of electrons; while the energy of the particles available for precise photographic recording diminished as the fourth power of the diameter of the canal tube (Aston 1924: 45). Above all, the parabolic images were diffuse and shadowed and unsuitable for accurate measurements because the particles were all moving with slightly different velocities. If only the rays could be focused!

The need for focusing positive rays led Aston (1919a) to the principle of the positive ray spectrograph or mass spectrograph, as Aston preferred to call the instrument, the idea for which was based upon an optical analogy which was justified theoretically by his Cambridge colleague, the mathematician R H Fowler. Just as white light is analysed into a spectrum of colours of different wavelengths by a prism, so will an electric field disperse a thin beam of positive rays whose individual particles are all moving with different velocities (Aston and Fowler 1922). A group of these rays moving with uniform velocity could then be filtered off through a fine slit and passed through a magnetic field, or magnetic prism, arranged at right angles to the electric field. This would bend 'the rays in the opposite direction to the foregoing electric field' (Aston 1924: 47). Rays of uniform mass would thus be focused as a 'mass spectrum' on a willemite screen or photographic plate, irrespective of their velocities. It will be understood, therefore, that the principal difference between Thomson's positive ray apparatus and Aston's spectrograph was that in the former the electric and magnetic fields were applied simultaneously at right angles; in the latter they were applied consecutively in the same plane. The resulting instrument surpassed the old apparatus in both resolving power and the intensity of the images it produced.

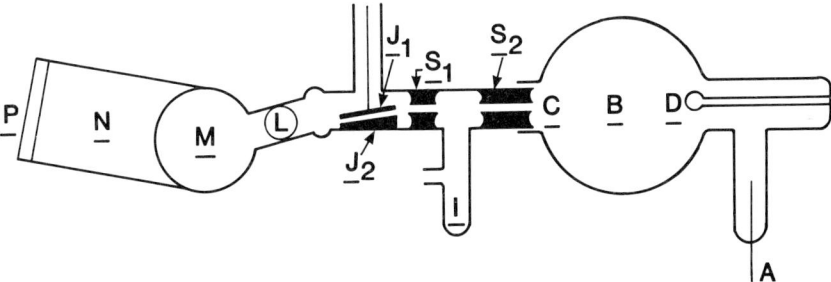

The principle of Aston's first mass spectrograph. An x-ray tube, B, is activated by an induction coil. A is the aluminium anode and C a concave aluminium cathode which is bored by the slit S_1, S_2, about 0.3 mm or less in diameter, which is kept exhausted by a charcoal tube, I. Cathode rays are spread by the electric plates J_1, J_2 at a potential of between 200 and 500 volts. Parts of the spectrum are then magnetically focused by a diaphragm, L, and the large magnet, M. The focused rays are recorded on a film plate, P, in the camera, N (from *Phil. Mag.* **38** (1919)).

Almost all the final engineering details of this instrument had to be done by trial and error. For example, finding the best position for the camera plate took days, while the fact that the slightest vacuum leak ruined the results meant that hours had to be spent discovering and sealing them. Nor was the actual experimental procedure less exacting. Only one photographic plate could be used a day since the film had to be loaded and dried overnight in conditions of vacuum. However, as many as six exposures could be contained on one plate. Even so, the rate of progress was clearly slow, a pace even more dictated by Aston's insistence on doing everything by himself.

The actual masses of the particles could be calculated empirically by several methods (Aston 1924: 57), the most obvious one being comparison with a calibration curve of reference lines of known mass-spectra such as $^6C^{2+}$, $^8O^{2+}$, ^{12}C, ^{16}O, ^{28}CO, $^{32}O_2$, $^{44}CO_2$. In this case, as long as the electric and magnetic fields remain constant, their values were not needed. In another method, that of 'coincidence', an unknown and known mass were again compared, but this time by so altering the fields that the unknown mass was focused at the same place as the known mass. The ratio m' to m was then given by $E'H'/EH$ where E' and H' are the values of the altered fields. In practice, since the strength of the magnetic field was difficult to determine accurately, it was kept constant and only E allowed to vary.

With his first instrument Aston (1919b) was able to show that neon was definitely a mixture of two isotopes; and within a few months he had analysed a sufficient number of other elements to be able to announce 'the whole number rule' to readers of *Nature* (Aston 1919c). According to this rule, if atomic masses were calculated with respect to $O = 16$, then with the exception of hydrogen, all masses were integral. (Incidentally, it was Aston (1920b) who introduced the isotopic symbolism, Ne^{22} etc, meaning neon of mass 22; for typographical reasons the symbolism was altered in the 1930s to ^{22}Ne, etc.) In a review of his own findings for *Science Progress*, Aston (1920a) concluded that Prout's hypothesis that atomic weights were whole numbers and multiples of the atomic weight of hydrogen had been vindicated; he also noted that his work supported Rutherford's model of the atom composed from only electrons and protons. It appeared, then, that the atomic weights of the elements which were determined by chemical methods were merely 'fortuitous statistical effects due to the relative quantities of the isotopic constituents' (Aston 1920b). In the example we have followed, neon consisted of two isotopes, ^{22}Ne and ^{20}Ne, whose relative natural abundance produced a non-integral chemical atomic weight of 20.2. Did it follow that the other aspect of Prout's hypothesis was also true? Were chemical elements really polymers of hydrogen, or rather of protons?

The problem here was that the mass of hydrogen was anomalous—it did not obey the whole number rule, for $H = 1.008$. However, as we saw earlier, Marignac (1860) had argued sixty years previously that the possible reason why Stas's very accurate determinations of atomic weights seemed to disprove Prout's hypothesis was that the individual masses of atoms in the aggregate were not strictly additive. Prout's law might be an 'ideal' law, like Boyle's law, which was subject to perturbing influences such that the masses of the subatomic particles of primordial matter did not add up exactly to the experimentally determined mass numbers. Later, as a consequence of his special theory of relativity, Einstein (1905) had concluded that mass could be conceived as a measure of the energy content of a body related by the familiar relationship, $e = mc^2$, where e is energy, m is mass and c is the constant velocity of light. One of the earliest theoretical applications of Einstein's equation was made by the Chicago chemist William D Harkins (1915), working in conjunction with a research student, E D Wilson (Siegel 1978). Harkins, who later beat Aston to the physical separation of chlorine isotopes (Harkins 1920), though not to their identification (Aston 1920b), was interested in the internal structure of atoms and in the possibility suggested by Rutherford's researches that they could be considered as nuclei built from α-particles or helium nuclei, surrounded by rings of electrons. The fact that helium masses, He = 4.003, did not add up to the determined atomic weights of other elements he explained by a packing together of the helium particles within the compound elementary atom. Citing Marignac's prescience (Harkins and Wilson 1915), he used Einstein's equation to show that the idea was tenable.

Aston (1921), who was no theoretician, seized on Harkins's packing concept in order to explain the anomaly of hydrogen; but using, of course, Prout's hydrogen model of the atom, not a helium one. Supposing elements were made from hydrogen atoms of mass 1.008. Mass was only additive, he argued, when nuclear charges were relatively distant from one another. Hence, in the case of atoms other than hydrogen, mass was lost within the atom itself in the form of binding energy as the individual atoms of hydrogen, the protons, were squeezed and packed together.

Thomson had from the first expressed serious doubts concerning Aston's technique and interpretations. Indeed, as late as March 1921, in a heated discussion at the Royal Society, Thomson (1921) confessed that he was still not completely convinced that hydride formation in the artificial conditions of the discharge tube might not explain Aston's results. Thus, in the case of chlorine isotopes, ^{35}Cl and ^{37}Cl, Thomson still felt that the latter might be $ClH_2 = 37$ (Thomson 1918). He was also sceptical of Aston's claim that his instrument was accurate to within 1 part in 10^3. However, once he had changed his mind, as he had to when Aston was awarded the Nobel Prize in 1922 'for his discovery by means of his mass spectrograph, of isotopes of a large number of non-radioactive elements, and for his enunciation of the whole number rule' (Aston 1966: ii), Thomson became a strong supporter. Indeed, Thomson and his Cavendish pupils implied subsequently in their later writings that he had always been Aston's constant inspiration (Aston 1942: 37). This conflict certainly soured their personal relationships and perhaps encouraged Aston in his proprietorial attitude towards 'his machine', for no one else in the laboratory was ever allowed to come near it. A shy and diffident bachelor, a superb experimentalist, but a poor theorist, he lacked the charisma of either Thomson or Rutherford. He detested lecturing and teaching, actually refusing to teach physics at Trinity College, where he held a permanent Fellowship. This fact, coupled with his seigniorial attitude towards his research and equipment explains why he had no pupils and founded no research school. Further, his independent wealth and delight in taking very long vacations around the world or skiing in Switzerland made him seem markedly different from the rest of the hardworking and less wealthy Cavendish community of the 1930s. As George Thomson (1946) remarked, 'Aston's mind was fundamentally that of an instrumentalist to whom experimental methods and actual manual dexterity are a joy in themselves approaching that to be gained from the results they give'.

Thomson's scepticism of Aston's work was not shared by the rest of the scientific community. Rutherford had no qualms, while Soddy was lavish in his praise. Aston's work, he declared—clearly paying off an old score in the rivalry between radiochemists and atomic physicists—was 'one of the most brilliant combinations of mathematical analysis and experimental skill that this century has produced. I do not think chemists have any reason to doubt the accuracy which he claims for his determinations' (Thomson 1921: 100).

As we shall see in the final chapter, Aston's work was of the greatest importance for theoretical physicists. It provided them with a remarkable tool for peering into the structure of atoms. For chemists it provided a new means for the accurate determination of the masses of elementary species. Finally, for geochemists and cosmologists it raised anew questions of how the elements had evolved and how and why they were distributed over the earth and the universe in the way observed. Why, to give one example of the sort of question raised by Aston's work, did elements of odd atomic number have no more than two stable isotopes?

As Aston (1924) worked steadily and laboriously through every element in the periodic table single-handedly searching for isotopes it became painfully clear to him that he had been over-hasty in announcing the whole number rule. It was not exact (Costa 1925). However, it soon became equally clear that the deviations from whole numbers could only be determined by building an even more accurate mass spectrograph and by using an especially sensitive photographic film ('Schumanised Plates') which he had developed with the cooperation of Ilfords (Aston 1923). By doubling the angles of the electric and magnetic deflections, and by sharpening the definition of the mass lines by using even finer slits, Aston succeeded triumphantly, and again single-handedly, in constructing a machine with five times more resolving power and an accuracy of 1 part in 10^4. This instrument was ready for use in 1927, and it immediately produced extensive evidence for deviations from the whole number rule (Aston 1927).

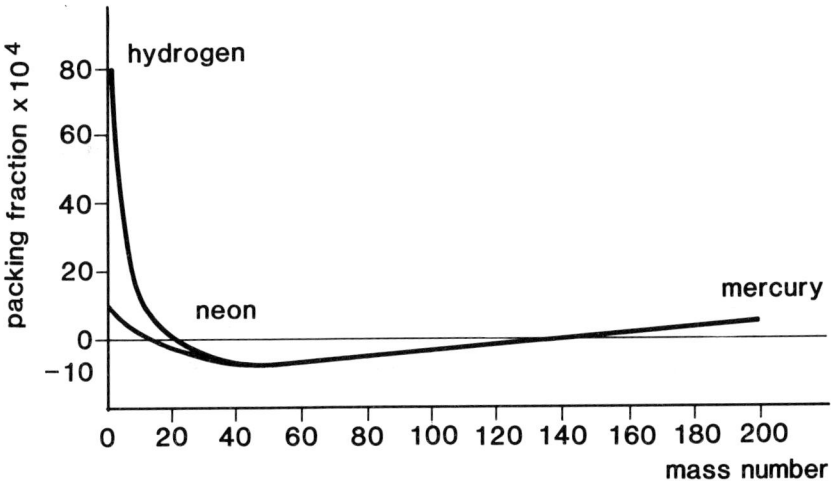

Aston's packing fraction. Positive mass defects are characteristic only of both the lightest and heaviest elements (from *Proc. R. Soc.* **115** (1927)).

These deviations were codified by extending his earlier notion of packing and by defining a new function called 'the packing fraction'. This function described the divergence of an atomic mass from a whole number divided by its mass number. By plotting packing fractions against mass numbers Aston (1927: 501) obtained a simple curve which threw important light on nuclear abundance and stability. Following Aston's lead, nuclear physicists came to regard the packing fraction, together with mass and atomic number, as one of the three defining constants of an atom, giving a measure of the binding force of the nucleus. A high packing fraction indicated loose packing and hence low stability; a low packing fraction indicated close integration and high stability. Equally importantly for the use of the mass spectrograph as a precision analytical instrument—the mass spectrometer—the packing fraction enabled different species of the same mass to be distinguished unambiguously (Panton 1981: 154).

The isotopic constitution of nearly all the elements in the periodic table was published by Aston in a constant stream of short communications to *Nature* between 1920 and 1935. Only a handful of elements, including hydrogen and oxygen, appeared to be anisotopic (i.e. not to possess isotopes). This being the case, Aston was extremely surprised to learn from Giauque and Johnston (1929) at the University of California, Berkeley, that two heavier isotopes of oxygen could be detected by ordinary spectroscopic means. It was some time before Aston was prepared to believe this; but in fact Aston's instrument was unreliable where isotopes are present in only minute amounts. ^{17}O and ^{18}O were also possibly masked in his machine by impurities such as ^{17}HO and ^{18}H$_2$O. As it turned out the band spectra of compounds proved to be a more reliable method of detecting new isotopes. Once aware of this, Aston asked C P Snow—a spectroscopist before he turned to literature and the civil service—to contribute two chapters on the isotopic effect in molecular spectra for the revised edition of *Isotopes* (Aston 1933a; 1942). Another consequence was that Aston (1931) had to urge the scientific community to accept the ^{16}O isotope as the standard unit for atomic weight values.

The discovery of the oxygen isotopes immediately suggested that hydrogen might also possess isotopes which Aston's two machines had failed to detect. Deuterium, ^{2}H, was indeed first detected spectroscopically in samples of heavy water by H C Urey and his collaborators in America (Urey *et al* 1932). At about the same time, at the University of Princeton, and working independently, Walter Bleakney (1932) not only confirmed the existence of heavy hydrogen with his newly constructed mass spectrometer, but also showed that what Thomson (1912), using the positive ray apparatus, had identified as H_3^+, was really hydrogen deutride, HD. Moreover, galled at not beating Urey to the discovery of deuterium (Panton 1981: 142), he made a determined search for a second heavier isotope of hydrogen. Bleakney's claim for the existence of tritium was made in 1935, but despite the concentration by electrolysis of 43 kilograms of Norwegian heavy water down to an 11 cm^3 sample, Aston (1937) was unable to verify the claim by mass spectrometry (Rutherford 1937; Eidinoff 1948). On re-examination Bleakney (1938) admitted that he had probably been misled by an impurity of mass 5, instead of the species required, DT$^+$, though a year later the existence of tritium was vindicated when Alvarez and Cornog (1939) prepared the radioactive gas by bombarding deuterium with itself. Its radioactivity had, of course, probably prevented it from being detected in the mass spectrometer.

These discoveries, while of great historical importance, were at the time a great nuisance to Aston, quite apart from the fact that they demonstrated that the field of isotopic research was no longer his alone. In practical terms it meant that Aston's estimation of the mass of hydrogen in 1927 was seriously in error and that the packing fractions for most of the elements had to be recalculated (Aston 1942: 121). Ironically, however, it had been the error in his published value for the mass of hydrogen which had suggested to Urey that an attempt to separate a 'heavy water' (if it existed) from ordinary water was theoretically a possibility (Urey 1932; Aston 1933b).

Aston in the Cavendish Laboratory with the third mass spectrograph c. 1937. Courtesy of the Cavendish Laboratory, University of Cambridge.

Aston (1935) was fond of saying that the goal of the experimental physicist was to 'make more, more and yet more experiments'. Not content, therefore, with the results he had obtained with his second mass spectrograph, and despite the fact that other scientists were building similar instruments—particularly in America—he decided in 1935 to construct an even more complex and accurate machine. His decision had much to do with his election to the Chairmanship of the International Committee on Atoms of the Union Internationale de Chimie, whose principal task was the compilation of an annual list of atomic weights and stable isotopes. A further stimulus came from the artificial disintegration of lithium and boron under the bombardment of protons by Cockcroft and Walton (1932) since this suggested that physicists would soon need to be able to detect the relations between the masses of transforming nuclei to four or five places of decimals (Aston 1942: 125). Unfortunately, Aston's third machine (Aston 1937), which was claimed to possess an accuracy of 1 part in 10^5, proved extremely difficult to adjust; and because of the intervention of war, Aston never succeeded in doing any significant work with it. By then, however, Aston's instruments had been surpassed in technical achievement and accuracy by a new breed of mass spectrometers developed by A J Dempster (1918), K T Bainbridge (1932) and A O Nier (1937), all of which Aston reviewed dispassionately in the final edition of his *Mass Spectra and Isotopes* (1942).

9

The Bottomless Pit

> Even if we resolve all matter into one kind, that kind will need explaining. And so on for ever and ever, deeper and deeper into the pit at whose bottom truth lies, without ever reaching it. For the pit is bottomless [O Heaviside 1893].

By the early 1930s, largely as a result of Aston's work, Prout's hypothesis that all atoms were made from a small number of building blocks had become, in Soddy's words, 'the corner-stone of modern theories of the structure of matter' (Soddy 1932). Rutherford's demonstration by the scattering of α-rays that atoms were largely empty space, but that their centres were occupied by a positively charged nucleus (Rutherford 1911; 1914; 1919), had led by the beginning of the First Word War to the planetary model of the atom. Because J J Thomson's work on positive rays indicated that no mass less than a positively charged hydrogen atom was ever found in the discharge tube, Rutherford (1920a; 1920b) took the logical step in identifying the hydrogen atom itself, or rather its nucleus, with positive electricity. Echoing Prout's *protyle* and, indeed, fortuitously Prout's own name, he suggested at the Cardiff meeting of the British Association for the Advancement of Science in 1920 that the particle should be called the 'proton'.

Rutherford's scattering experiments had also revealed that the positive charge on the nucleus was approximately one half of the atomic weight—a number which the Dutch schoolteacher van den Broek (1914) noticed was the same as that of an element's place in a series arranged in order of increasing atomic weight. He suggested that this 'atomic number' was equal to the positive charge on the nucleus, a speculation which was soon confirmed by the differences in wavelengths of the characteristic x-rays given off by elements when bombarded by cathode rays in discharge tubes (Aston 1924:111).

Barkla (1911) had previously shown that the number of electrons associated with each element was also roughly half of the atomic weight, and that the β-rays (electrons) produced by radioactive elements seemed to originate from what Rutherford described as the nucleus. The planetary model was therefore complete: a nucleus of charge A, composed from 2A protons and A electrons (thus producing the overall positive charge, A) surrounded by A electrons to make the complete system electrically neutral. The loss and gain of electrons could then be used to explain chemical combination.

Bohr (1913) then argued that if the outer planetary electrons rotated rapidly round the nucleus, as charged particles they would give off electromagnetic rays which would eventually lead them to spiral towards and collapse into the nucleus. To prevent this, he seized upon Planck's principle that radiation was not emitted continuously but in packets or quanta of energy. If electrons were postulated to rotate in fixed orbits around the nucleus, the quanta emitted as they jumped from one orbit to another could be correlated precisely with the spectra of simple elements such as hydrogen and helium.

Although elementary chemistry was taught with a strongly historical basis in British schools until the 1960s and Prout's hypothesis would have been familiar to members of the British scientific community who followed a Lower School Certificate or Ordinary Level chemistry syllabus between 1930 and 1960, this was not the case with continental and American scientists. In 1949 Imperial Chemical Industries could use a drawing of Prout in a series of newspaper advertisements confident in the knowledge that 'Prout's hypothesis' would strike a chord in the minds of educated readers; no comparable use of Prout's image can be found in the American press. Insofar as matter theory became the province of physicists and research on the nature of matter became much more of an international activity in the 1930s, it is perhaps scarcely surprising that the historian finds few explicit references to Prout in the literature of post-war particle physics. A few scientists, such as Aston (1942), Witner (1946) and Fowler (1964), who were clearly conscious of the historical and intellectual tradition they were following, do mention Prout; but the majority have been unconscious of the Proutian nature of their obsession for simplification. What follows, therefore, by way of dénouement, is an historian's interpretation of what has happened in particle physics since 1930.

The snag with the simple Proutian model of protons and electrons as perfected by Bohr was (as both Rutherford and Aston recognised) that the nucleus had to consist of a combination between 'the slow and ponderous proton with the light and lively electron' (Rutherford

1940). In a discussion at the Solvay conference on atoms and electrons in 1921 (Mehra 1975:103), Jean Perrin speculated that the source of solar energy probably lay in the transformation of hydrogen into helium, with the loss of mass from packing being released as solar heat and light. Rutherford agreed, adding that 'it has occurred to me that the hydrogen of the nebulae might consist of particles which one might call the "neutrons", which would be formed by a positive nucleus with an electron at a very short distance'. Although Rutherford at this time chiefly saw the hypothetical neutron as an explanation of how positively charged particles 'could penetrate into nuclei against the forces of repulsion' (Mehra 1975:103; Rutherford 1920b), no experimental evidence for such a composite particle was forthcoming until 1932. It was only then that Rutherford's former pupil, James Chadwick (1932a; 1932b) detected such a neutral particle by the ionisation of particles recoiling from their impact.

The discovery of a third atomic particle, the neutron, still preserved the simplicity of the Proutian atom; indeed, it seemed to enhance it. All atomic nuclei could now be supposed to consist of a combination of protons and neutrons. For example, the oxygen isotope mass 16, had 8 protons and 8 neutrons, the isotope, mass 17, had 8 protons and 9 neutrons. On the other hand, by 1932, developments in the theoretical domain were already undermining the simplicity and elegance of the 'tria prima' of electrons, protons and neutrons. If the electrons in the nucleus were tied to protons as 'neutrons', where did the β-rays of radioactive disintegrations come from? And why, in certain artificial transmutations and cosmic-ray scatterings noted by Anderson (1932; Hanson 1963), was a positive electron or 'positron' formed? Theoreticians produced answers, but only at the cost of 'inventing' more particles.

Thus, Enrico Fermi (Wilson 1968), working in Rome, satisfactorily explained the phenomenon of the β-decay of radioactive elements in 1934 by accepting the existence of the 'neutrino'—a virtually massless neutral particle which had been postulated three years earlier by Wolfgang Pauli (Mehra 1975: 226) in order to preserve the principle of conservation of energy. According to Fermi and Pauli, nuclear neutrons decayed into protons and electrons (the β-rays), together with a neutrino

$$n = p + e^- + \bar{v}.$$

Here, the price of preserving one of the grandest simplifying assumptions of nineteenth-century physics—conservation of energy—was an increase in the number of basic particles. In Fermi's theory electrons were not pre-existent, or confined, within neutrons, but produced 'at the moment when they are emitted'. In fact neutrinos remained undetected until 1956 when they emerged in their millions from the Savannah River nuclear reactor in America (Reines and Cowan 1956).

Again, during the 1930s, interest in ionising cosmic-ray showers in the earth's atmosphere and their ability to penetrate deeply into the earth's surface—phenomena which could not be explained in terms of either showers of protons or neutrons—forced physicists to invent further heavy particles called 'mesitrons' or 'mesons' (Anderson and Neddermeyer 1938). Fortuitously, these mysterious particles seemed to fit in well with a theory of how protons and neutrons bound themselves together to form atomic nuclei by the exchange of what the Japanese physicist Yukawa (1935; Yukawa and Kikuchi 1950) had called 'heavy quanta' or 'yukons'. In 1938 Yukawa took the brave step of suggesting that the nuclear exchange force (now known as the strong force) was caused by the mediation of mesons (Snyder 1950), the term most widely adopted after World War Two for cosmic ray particles. (In fact, Yukawa's strong force is mediated by the μ-meson, or muon.) Hence, although on the eve of war the simple planetary atom of protons, neutrons and electrons had given way to one which involved invisible neutrinos and photons and ponderous mesons, with the exception of the anomalous positron, all the particles correlated theoretically with the observed properties of atoms.

Time and deeper reflection showed, however, that the meson (muon) could not possibly be either Yukawa's strong force mediator, or (since it was unstable and decayed into an electron and neutrino) the original source of particle energy entering the earth's atmosphere as cosmic rays. Immediately following the war, therefore, Powell, at the University of Bristol, made an intensive study of cosmic rays. In 1947 he detected the very heavy π-meson (pion) and showed that it decayed into the muon, which in its turn produced an electron plus two neutrinos (Snyder 1950: 45). Theoretical work by Marshak and Bethe (1947) and Sakata and Inowe (1946) suggested that these pions, not the muons, were responsible for the nuclear force. This proved to be the case, despite the complication of the discovery in cyclotrons of a third meson, the kaon (Leprince-Ringuet 1949). All three mesons were found to exist in positive, negative and neutral states and each required its own particular kind of neutrino. In the 1950s further heavy mesons, lamda (Λ) sigma (Σ), xi (Ξ) and delta (Δ) were identified and each was to be found (or was expected to exist) in charged forms called resonances. Thus, by 1953, physicists were aware of some twenty forms of subatomic matter—a number which could be approximately doubled when their anti-particle opposites, which appeared in some collision experiments, were included in the accounting.

With the advent of high energy accelerators in the 1950s and 1960s, which allowed accelerated protons and other particles to be fired at chosen targets, it seemed likely that yet further strange entities would be found in the debris of photographed collisions and scatterings. Once more the scientific community was in a similar position that mid-nineteenth-century chemists had confronted with the elements. With the simplicity of the nuclear atom of the 1930s gone for ever, physicists seemed to have invented a Catherine-wheel in their expensive accelerators which threw out streams of jigsaw pieces faster than the theoretician could detect any pattern. Indeed, when the *Review of Modern Physics* published a list of observed particles for the last time in 1976 (thereafter the listing has appeared intermittently in monograph form) it occupied 245 pages (Trefil 1980: 112).

Nevertheless, as Lucretius had observed centuries before:

> However endlessly infinite the Universe may be, yet the smallest things will equally consist of an infinite number of parts. Since true reason cries against this and denies that the mind can believe it, you must needs give in and admit that there are least parts which themselves are partless [Lucretius 56BC].

Although natural theology was no longer to be invoked (though many physicists invoked a Platonic God of Mathematics) physicists instinctively reasoned that matter must ultimately be simple (Whyte 1950–1). The principle of parsimony—the preference for theories demanding fewer assumptions and giving simple and preferably quantitative explanations—therefore drove them to search for a Prout's hypothesis of the nucleus.

The obvious way to do this was to follow the path of nineteenth-century chemists who, through natural history and the careful cataloguing of the properties of elements, had discovered that they fell into natural groups or families. This periodicity, it will be recalled, then suggested an underlying developmental or evolutionary pattern (Crookes 1886) whose details were finally unravelled by Aston in the 1920s. There was, however, an alternative to reducing the number of elementary particles, which again had historical precedents, namely to abandon the notion of elementary particles completely (Boyle 1661; Brodie 1867; Chew 1964; 1970).

In practice, of course, the classification of particles had been in progress throughout the 1950s. For example, by the separation of leptons (weak force interactors like electrons, muons and neutrinos) from hadrons (the strong force interactors like protons and neutrons); by classifying into bosons (particles with an integral spin which obeyed Bose statistics) and fermions (particles with half-integral spins which obeyed Pauli's exclusion principle and Fermi statistics); by distinguishing baryons ('heavy ones'), which always decay eventually into protons, from mesons, which always decay into leptons and photons, the number of protons produced from a baryon being called 'the baryon number'; by differentiating between the rates of decay as comparatively sluggish, 10^{-10} seconds ('strange particles') and fast, 10^{-23} seconds (non-strange particles) and introducing a new conservation principle called 'strangeness' (Trefil 1980). Although these methods of sorting elementary particles, together with older ones like electric charge which were familiar from quantum mechanics, did not in themselves produce simplifications (why, for example, should there appear to be many more hadrons than leptons?), they did draw attention to the integral or fractional (quantum) numbers associated with each kind of particle: spin, charge, baryon number, strangeness and, for hadrons, isotopic spin, etc. It was amongst some of these numbers that Murray Gell-Mann and Yuval Ne'eman (1961) detected a significant pattern in 1961 when they were independently trying to update Yukawa's theory of strong interaction (Pickering 1984: 57). Gell-Mann, who has become the dominant figure in dictating the exotic vocabulary of high energy physics since 1953, when he coined the term 'strangeness', called this pattern the 'eightfold way'—a reference to the Buddhist path of enlightenment. The key to this grouping, of which the family of eight echoing Mendeleeff's octaves of elements in the periodic table was but one of several groupings, lay in an abstract mathematical theory of rotational symmetry called SU(3), 'special unitary group in three dimensions'. This theory enabled particles which possessed the same spin and parity, but different isospin and strangeness, to be grouped together in a 'graph' of strangeness (or, more technically, hypercharge = baryon number + strangeness) plotted against isospin.

For example, ignoring for our purposes any exact scale, Gell-Mann was able to array a quartet of Δ (delta) particles, a triplet of Σ^* and a doublet of Ξ^* into the following pattern (Polkinghorne 1979: 60; Pickering 1984: 59).

$$\Delta^- \quad\quad \Delta^0 \quad\quad \Delta^+ \quad\quad \Delta^{++}$$
$$\Sigma^{*-} \quad\quad \Sigma^{*0} \quad\quad \Sigma^{*+}$$
$$\Xi^{*-} \quad\quad \Xi^{*0}$$
$$?$$

As Polkinghorne (1979: 61) has commented, it does not take a child of very advanced age to suggest that there ought to be a tenth particle, ?, in this array to complete the triangular symmetry. Both Gell-Mann and Ne'eman took this step and, like Mendeleeff predicting the existence and properties of gallium and scandium from the periodic table of the elements, they predicted, to an amazing degree of accuracy, the mass and appropriate quantum numbers of an unknown sigma minus (Λ^-) particle. Their predictions were completely fulfilled when this entity was detected in the Brookhaven Alternating Gradient Synchrotron in 1964, and both men were duly awarded the Nobel Prize for physics in 1969.

It should not be thought that the demonstration of the eightfold way and its apparent vindication with Λ^- was met with universal support by the high energy physics community in the 1960s (Pickering 1984). To many critics it seemed no more than a heuristic device, whilst to others, like the American, Geoffrey Chew (1964; 1970), it led to a questioning examination of the very concept of elementary particles (Schrader-Frechette 1977). Since, Chew reasoned, much, if not most, of the physics of hadron interactions was based solely on the observation of particles before and after scatterings and collisions, all that was 'supposed' to happen in between by way of entities like pions and neutrinos, which were constructed by physicists in order to save conservation laws, did not necessarily have any physical reality. All that could really be said was 'enshrined in the set of transition probabilities that from given initial states one will arrive at given final states' (Pickering 1984: 73). Such probabilities involved a branch of mathematics called the 'scattering-' or S-matrix, from which Chew fashioned a 'bootstrap' philosophy in which all hadrons were generated from one equation and there were no fundamental entities: paradoxically, everything was made out of everything else, as Leibniz had suggested in the seventeenth century (Gale 1974). Each identified entity owed its existence to forces to which the entity itself contributed; there were no truly 'elementary particles', for nuclear interactions were merely an interacting system possessing particular properties.

In direct contrast to Chew's bootstrap model, Gell-Mann (1964) and George Zweig (1964) independently reaffirmed belief in the existence of fundamental entities. Taking the Proutian route they suggested that the symmetries of the 'eightfold way' could be explained very simply by positing the existence of three ultimate entities called 'quarks'. (Zweig's term 'aces' could not compete with Gell-Mann's allusion to James Joyce's *Finnegans Wake*.) Two quarks ('up' and 'down') generated all hadrons with zero strangeness, while all strange hadrons contained one or more 'strange' quarks. Moreover, when three 'anti-quarks' were also admitted, all the known mesons could also be generated. Although the hypothesis elegantly generated nucleons, there were too many difficulties in the way for the quark theory to be acceptable before the 1970s. A major stumbling block was that quarks had to possess non-integral electrical charges of $+2/3$ (up), $-1/3$ (down) and $-1/3$ (strange)—the antiquarks had correspondingly opposite charges; yet protons and electrons had integral charges. On the other hand, this did mean that physicists had an unequivocal tool for quark detection—fractional electrical charge. Unfortunately, despite a decade of searching for such free particles in cosmic rays, accelerators and in repetitions of Robert Millikan's classic oil-drop experiment on the charge of an electron, absolutely no evidence for the free existence of quarks was forthcoming (Polkinghorne 1979: 65; Pickering 1984: 88).

Although this failure reinforced Chew's suspicion that quarks were merely mathematical artefacts, scattering experiments in 1968 based directly on the analogy of Rutherford's experiment which defined the nucleus in 1911, provided unequivocal evidence that quarks were 'trapped' or 'confined' (Nambu 1976) within protons and neutrons. Thus, by the early 1970s, some evidence for the existence of real quarks was forthcoming. On the other hand, to explain their confinement, physicists had to posit a quark force in which entities called 'gluons' were exchanged between quarks, just as pions were exchanged between protons and neutrons to produce the nuclear strong force. That, once again, physicists considered it worth multiplying entities was due to another awkward feature of the quark model.

In Gell-Mann's model the quartet of non-strange delta particles, Δ^-, Δ^0, Δ^+ and Δ^{++}, was composed from up and down quarks as follows (Polkinghorne 1979: 102; Close 1983: 72):

	Sum of quark charges	Quark cluster
Δ^-	-1	ddd
Δ^0	0	udd
Δ^+	$+1$	uud
Δ^{++}	$+2$	uuu

Similarly, strange hadrons, like the Ξ quartet and the Λ^- particle, involved a third 'strange' quark, the Λ^- being composed from three identical strange quarks. Now, according to Pauli's exclusion principle, which had always enjoyed success in explaining the electronic orbits of the elements, two identical quarks with the same spin should not occupy the same state. Three was inconceivable. To preserve Pauli's principle, therefore, it had to be supposed that the three quarks possessed some further property which distinguished them—a quality which Oscar Greenberg called 'colour' (Pickering 1984: 216–22). However, Greenberg's solution, and its extension by the

Japanese physicist, Yoichiro Nambu, which posited nine varieties of quark, remained obscure to most physicists until the early 1970s (Pickering 1984: 218) when deeper studies of pion decay routes began to favour the existence of coloured quarks.

A further complication of the quark model came in 1974 when two widely separated teams of particle physicists detected the ψ and J particles—named the ψJ particle after it was realised they were the same object (Trefil 1980: 173; Pickering 1984: 253). Now Gell-Mann had already speculated on the existence of a fourth quark to form a pair with the third strange quark, and this idea was developed further by S L Glashow and his colleagues in 1964 in order to explain certain decay anomalies. It was Glashow who christened the new quark 'charm', but its existence had remained speculative until the ψJ particle was discovered. This discovery, and the predictions of the existence of further heavy particles which now followed from the confirmation of the strength of a four-quark model, seemed to be the vindication of Gell-Mann's particulate model over Chew's bootstrap nihilism.

Hence, by 1974, with the addition of colour and charm, there were twelve quarks (together, of course, with their anti-quark mirror images). Nor, in the last ten years, has the number of basic building blocks decreased. It soon became necessary to introduce a fifth 'top' or 'truth' quark as evidence accumulated for the existence of a new heavy lepton, the upsilon, Y (Lederman 1978). This, in turn, implied the existence of its partner, a sixth 'bottom' or 'beauty' quark, which was eventually detected in 1984.

With six quarks in three colours—an array of eighteen fundamental nuclear entities—together with the six leptons (the electron, muon, taon and their three companion neutrinos), making twelve or twenty-four ultimate entities according to one's system of accounting, the success of the Proutian enterprise may be doubted. However, this is not how it is seen by the physics' community. Six flavoured quarks neatly balances six leptons; lepton–neutrino pairing is balanced by the pairing of quarks, ud, cs, tb; and just as taons and muons seem to be heavier versions of the electron, so do top/bottom and charm/strange quarks appear to be generators of second- and third-generation hadrons (Miller 1984). Moreover, theoretical physicists have begun to be successful in bringing together the various fields of force which bind quarks and leptons and they are confident that the quark model will eventually lead to a 'unified field theory' in which gravitational, strong, weak and electrodynamic forces which mediate or affect aggregates of hadrons and leptons will be summarised in a few elegant equations.

Bibliography

Archival materials are abbreviated as follows:
  Ballogie papers: in private possession;
  Sidmouth: Lockyer's Sidmouth Observatory papers, University of Exeter;
  Wellcome Institute Library: Wellcome Institute for the History of Medicine, London;
  Royal Inst. Cornwall: Library of the Royal Institution of Cornwall, Truro.

Abney W and Festing E R 1887 On the influence of the atomic grouping in the molecules of organic bodies on their absorption in the infra-red region of the spectrum *Phil. Trans. R. Soc.* **172** 887–918
Allen W and Pepys W H 1808 On the changes produced in atmospheric air and oxygen gas by respiration *Phil. Trans. R. Soc.* 249–81
Alvarez L W and Cornog R 1939 Helium and hydrogen of mass 3 *Phys. Rev.* **56** 613
Ampère A M 1815 On the determination of the proportions in which substances combine *Phil. Mag.* **45** 41–3, 109–16, 188–93, 344–9
—— 1820 Sur l'action des courans voltaiques *Ann. Chim.* **15** 59–76, 170–218
Anderson C D 1932 The positive electron *Phys. Rev.* **43** 491–4
Anderson C D and Neddermeyer S M 1938 Cosmic-ray particles of intermediate mass *Phys. Rev.* **54** 88–9
Annals of Philosophy 1813 Prout's Lectures **2** 312
—— 1814 Prout's Lectures **3** 151
Anon 1820a The death of John Prout *Gents' Mag.* 284 (September)
—— 1820b Review of progress in physiology *Ann. Phil.* **16** 113
—— 1831 The Bridgewater bequest *Literary Gazette* 5 February
—— 1832 Criticism of Prout *Dublin J. Med. Sci.* **1** 66
—— 1841 Review of Prout, *On Stomach, Brit. For. Med. Rev.* **11** 336
—— 1843 Review of Prout. *On Stomach, Brit. For. Med. Rev.* **16** 478
—— 1846 Review of Prout's Bridgewater treatise *Brit. For. Med. Rev.* **21** 123
—— 1848a The Rev John Turner *Gents' Mag.* **ii** 215

—— 1848b Review of Prout, *On Stomach, Dublin J. Med. Sci.* **6** 168
—— 1850 William Prout *Med. Times* **1** 17
—— 1851 Dr. William Prout *Edinb. Med. Surg. J.* **76** 126–83
—— 1880 Recent science *Nineteenth Century* **7** 521–7
Apjohn J 1822 On the influence of moisture in modifying the specific gravity of gases *Ann. Phil.* **19** 385–7
Armstrong H E 1878 The nature of the elements *Chem. News* **38** 291–2
—— 1879 The dissociation of chlorine *Nature* **20** 357–8
—— 1928 Norman Lockyer's work and influence *Nature* **122** 870–4
Aston F W 1907 Experiments on the length of the cathode dark space *Proc. R. Soc.* A **79** 80–95
—— 1908 Experiments on a new cathode dark space in helium and hydrogen *Proc. R. Soc.* A **80** 45–9
—— 1912 Sir J J Thomson's new method of chemical analysis *Sci. Prog.* **7** 48–65
—— 1913 A new elementary constituent of the atmosphere *Br. Assoc. Rep.* 403
—— 1919a A positive ray spectrograph *Phil. Mag.* **38** 707–14
—— 1919b Neon *Nature* **104** 334
—— 1919c The constitution of the elements *Nature* **104** 393
—— 1920a Mass spectra and the atomic weights of the elements *Sci. Prog.* **15** 212–22
—— 1920b The mass spectra of the chemical elements *Phil. Mag.* **39** 611–25
—— 1921 Isotopes and atomic weights *Proc. R. Inst.* **23** 299–310
—— 1923 The mass-spectra of chemical elements (Schumanised photographic plates) *Phil. Mag.* **45** 934–45
—— 1924 *Isotopes* 2nd edn (London: Edward Arnold)
—— 1927 A new mass-spectrograph and the whole number rule *Proc. R. Soc.* **115** 487–514
—— 1931 Discussion on the unit of atomic weight *Br. Assoc. Rep.* 333
—— 1933a *Mass Spectra and Isotopes* 3rd edn (London: Edward Arnold)
—— 1933b The hydrogen isotope of mass 2 *Sci. Prog.* **28** 203–5
—— 1935 The story of isotopes *Br. Assoc. Rep.* 23–4
—— 1937 A second order focussing mass spectrograph and isotopic weights by the doublet method *Proc. R. Soc.* A **163** 391–404
—— 1942 *Mass Spectra and Isotopes* 4th edn (London: Edward Arnold)
—— 1966 Mass spectra and isotopes (Nobel Prize Lecture 1922) (in Nobel (1966) **ii** 7–20)
Aston F W and Fowler W 1922 Some problems of the mass spectrograph *Phil. Mag.* **43** 514–28
Athenaeum 1834 Review of Prout's Bridgewater treatise **58** 349
Avogadro A 1811 Essay on the determination of the relative masses of elementary molecules in *Foundations of the Molecular Theory* (Alembic Club Reprint no 4, Edinburgh 1950)
—— 1814 Mémoire sur les masses relatives des corps simples *J. Physique* **78** 131–56
Badash L 1969 *Rutherford and Boltwood Letters on Radioactivity* (New Haven, CT: Yale University Press)

Badash L 1979a The suicidal success of radiochemistry *Br. J. Hist. Sci.* **12** 245–56
—— 1979b *Radioactivity in America* (Baltimore: Johns Hopkins Press)
Bainbridge K T 1932 A mass spectrograph *Phys. Rev.* **40** 130
Barkla C G 1911 The distribution of secondary X-rays *Phil Mag.* **21** 270–8; **22** 396–412
Beaumont W 1833 *Experiments and Observations on Gastric Juice and the Physiology of Digestion* (reprint 1959 New York, Dover Books)
Bécquerel H 1896 Sur les radiations invisibles émises par les corps phosphorescents *C. R. Acad. Sci., Paris* **122** 420–1, 1086–8 (Engl. Transl. Romer (1964))
Benfey O T 1952 Prout's hypothesis *J. Chem. Ed.* **29** 78–81
Benton E 1974 Vitalism in 19th century thought *Stud. Hist. Phil. Sci.* **5** 17–48
Berthelot M 1855 *Les Origines de l'Alchimie* (Paris: Georges Steinheil)
Berzelius J 1813a Essay on the cause of chemical proportions *Ann. Phil.* **2** 443–54
—— 1813b On the composition of animal bodies *Ann. Phil.* **2** 19–26, 195–208, 377–87, 415–25
—— 1814a On the cause of chemical proportions *Ann. Phil.* **3** 51–62, 93–106, 244–55, 353–64
—— 1814b Experiments to determine the definite proportions in which the elements of organic nature are combined *Ann. Phil.* **4** 323–31, 400–9
—— 1819 *Essai sur la Théorie des Proportions Chimiques* (Paris: Mequignon-Marvis)
—— 1823 On the multiples hypothesis *Jahres Ber.* **2** 39–40
—— 1825 Review of Prout (1822b) *Jahres Ber.* **4** 239–45
—— 1825–31 *Lehrbuch der Chemie* 4 vols (Dresden: Arnold)
—— 1827 Review of Prout's hypothesis *Jahres Ber.* **8** 244
—— 1829–33 *Traité de Chimie* 8 vols (Paris: Firmin Didot)
—— 1844 On the atomic weight of carbon *Jahres Ber.* **23** 26–30
—— 1845 On the hypothesis of Dr Prout with regard to atomic weights *Silliman's Am. J. Sci.* **48** 369–72
—— 1845–50 *Traité de Chimie* 6 vols (Paris: Firmin Didot)
—— 1903 *Jac. Berzelius Reseanteckningar* ed H G Söderbaum 2 vols (Stockholm: Kungliga Svenska Vetenskaps Akademien)
—— 1912–36 *Jac. Berzelius Bref* ed H G Söderbaum 6 vols (Stockholm/Uppsala: Almqvist and Wiksell)
Berzelius J and Dulong P L 1821 New determinations of the proportions of water and of specific gravities *Ann. Phil.* **18** 48–50
Bichat X 1805 *Recherches Physiologiques sur la Vie et la Mort* 3rd edn (Paris: Brosson)
Bidder H F and Schmidt C 1852 *Die Verdauungsaefte und der Stoffwechsel* (Beyer: Mitau)
Bird G 1844 *Urinary Deposits, their Diagnosis, Pathology and Therapeutical Indications* (London: Churchill)
Blane Sir G 1825 *Elements of Medical Logic* (London: T and G Underwood)
Bleakney W 1932 Additional evidence for an isotope of hydrogen mass 2 *Phys. Rev.* **39** 536; **41** 32–8

—— 1935 Concentration of tritium (H^3) *J. Am. Chem. Soc.* **57** 780
—— 1938 On the existence of H^3 *Phys. Rev.* **54** 388
Blumenbach J F 1815 *The Institutions of Physiology* transl. J Elliotson (London: Bensley)
Bohr N 1913 On the constitution of atoms and molecules *Phil. Mag.* **26** 1–25
Bolton H C 1898 The revival of alchemy in France *Chem. News* **77** 74
Boltwood B B 1907 Note on a new radio-active element [ionium] *Am. J. Sci.* **24** 370–2
Boyle R 1661 *The Sceptical Chymist* (reprinted 1911 etc Everyman Library, London: Dent)
Braconnot H 1836 Expériences chimiques sur le suc gastrique *Ann. Chim.* **59** 348–58
Brande W T 1808 On the differences in the structure of calculi *Phil. Trans. R. Soc.* 223–43
—— 1818 Review of Thomson, *System of Chemistry, J. Sci. Arts* **4** 299–321
—— 1823 Table of representative numbers *Q. J. Sci.* **14** 49–63
—— 1827 Review of work of E. Turner *Q. J. Sci.* [2]**2** 60–7
—— 1848 *A Manual of Chemistry* 6th edn 2 vols (London: J W Parker)
Brayley E W 1832 On the true source of the amniotic acid of Vauquelin *Phil. Mag.* [3] **1** 319–20
British Association Reports 1832 *Report for the Year 1831*
—— 1835 *Report for the Year 1835*
—— 1837 *Report for the Year 1837*
Brock W H 1963 Prout's chemical Bridgewater treatise *J. Chem. Ed.* **40** 652–5
—— 1964 Who were the editors of the *Annals of Medicine and Surgery? Med. Hist.* **8** 291–3
—— 1966 The selection of the authors of the Bridgewater treatises *Notes Rec. R. Soc.* **21** 162–79
—— 1967 Prout on taste, smell and flavor *J. Hist. Med.* **22** 184–7
—— 1970 William Prout on barometry *Notes Rec. R. Soc.* **24** 281–94
Brock W H and Knight D M 1965 The atomic debates *ISIS* **56** 5–25
Brock W H, McMillan N D and Mollan R C 1981 *John Tyndall* (Dublin: Royal Dublin Society)
Brodie B C 1867 Ideal chemistry *Chem. News* **15** 295–305
—— 1879a On the decomposition of the simple weight χ *J. Chem. Soc.* **35** 673–82
—— 1879b *Letter to Lockyer* 8 July (Sidmouth)
van den Broek A 1914 On nuclear electrons *Phil. Mag.* **27** 455
Brooke J H 1968 Wöhler's urea and its vital force? A verdict from the chemists *Ambix* **15** 84–114
—— 1971 Organic synthesis and the unification of chemistry—a reappraisal *Br. J. Hist. Sci.* **5** 363–92
—— 1978 The natural theology of the geologists: some theological strata in *Images of the Earth: Essays in the History of the Environmental Sciences* (Chalfont St Giles: Br. Soc. Hist. Sci.) 39–64
Brown S 1846 Experimental researches on the production of silicon from paracyanogen *Trans. R. Soc. Edinb.* **15** 165–76, 229–46

Buchdahl G 1959 Sources of scepticism in atomic theory *Br. J. Phil. Sci.* **10** 120–34
Buckland W 1829 On the discovery of coprolites, or fossil faeces *Trans. Geol. Soc.* **3** 223–38 [Prout's analyses 237–8]
Bud R 1980 The discipline of chemistry. The origins and early years of the Chemical Society of London *PhD Thesis* University of Pennsylvania
Burchfield J E 1975 *Lord Kelvin and the Age of the Earth* (London: Macmillan)
Cannizzaro S 1858 *Sketch of a Course of Chemical Philosophy* (Alembic Club Reprint no 18, Edinburgh)
Cannon W [later S F] 1960 The uniformitarian-catastrophist debates *ISIS* **51** 38–55
—— 1960-1 The problem of miracles in the 1830s *Victorian Stud.* **4** 5–32
Cantor G N and Hodge M J S 1981 *Conceptions of Ether. Studies in the History of Ether Theories 1740–1900* (Cambridge: Cambridge University Press)
Cardwell D S L 1968 *John Dalton and the Progress of Science* (Manchester: Manchester University Press)
Carnelley T 1886 Suggestions as to the cause of the periodic law and the nature of the chemical elements *Chem. News* **53** 157 (reproduced in Knight (1970))
Chadwick J 1932a The possible existence of a neutron *Nature* **129** 312
—— 1932b The existence of the neutron *Proc. R. Soc.* A **136** 692–708
Chambers R 1844 *Vestiges of the Natural History of Creation* (London: John Churchill)
Chapman C B 1967 Edward Smith *J. Hist. Med.* **22** 1–26
Chenevix R 1803 Enquiries concerning . . . a new metallic substance [palladium] *Phil. Trans.* 290–320
Chevreul M E 1824 *Considérations Générales sur l'Analyse Organique* (Paris: F G Levrault)
Chew G F 1964 Elementary particles? *Phys. Today* **17** 30–4
—— 1970 Hadronic bootstrap: triumph or frustration *Phys. Today* **23** 23–7
Children J G 1824 On the nature of the free acid ejected from the human stomach in dyspepsia *Ann. Phil.* **24** 68–9
—— 1825 A summary view of the atomic theory according to the hypothesis adopted by M. Berzelius *Ann. Phil.* **25** 185–93
Christison R 1837 *Biographical Sketch of the late Edward Turner* (Edinburgh: J Carfrae)
Clarke F W 1873 Evolution and the spectroscope *Pop. Sci. Mon.* (1) January 520–6
Close F 1983 *The Cosmic Onion. Quarks and the Nature of the Universe* (London: Heinemann)
Cockcroft J D and Walton E T S 1932 The disintegration of elements by high velocity protons *Proc. R. Soc.* A **136** 229–42
Coindet J 1820 Découverte d'un nouveau remède contre le goître *Ann. Chim.* **15** 49–59
Cole F J 1921 History of anatomical injection (in Singer (1921))
Coley N G 1971 Animal chemists and the urinary stone *Ambix* **18** 69–93
Costa J L 1925 Mass spectra of some light elements *Ann. Phys., Paris* **4** 425
Crookes W 1861 The composition of the solar spectrum *Chem. News* **4** 293

—— 1878a Repulsion resulting from radiation *Phil. Trans. R. Soc.* **169** 243–318

—— 1878b *Letter to Lockyer* 16 December (Sidmouth)

—— 1884 On radiant matter spectroscopy *Phil. Trans. R. Soc.* **174** 891–918

—— 1886 On the nature and origin of the so-called elements *Br. Assoc. Rep.* 558–76

—— 1887 Review of Lockyer's *Chemistry of the Sun, Chem. News* **56** 267

—— 1888 Elements and meta-elements *J. Chem. Soc.* **53** 487–504

—— 1902 The stratifications of hydrogen *Proc. R. Soc.* **69** 399–413

—— 1912 Address in *Presidential Addresses to the Society for Psychical Research* (London: Society for Psychical Research) 86–103

Curie M 1898 Rayons émis par les composes de l'uranium et du thorium *C. R. Acad. Sci., Paris* **126** 1101–3; **127** 175–8

Curie P and Curie M 1898 Sur une substance nouvelle radioactive contenue dans la pechblende *C. R. Acad. Sci., Paris* **127** 175–8

D'Albe E E F 1923 *The Life and Work of Sir William Crookes* (London: Fisher Unwin)

Dalton J 1808–10 *A New System of Chemical Philosophy* (Manchester: R Bickerstaff)

Daniell J F 1843 *An Introduction to the Study of Chemical Philosophy* 2nd edn (London: J W Parker)

Darwin C 1887 *Life and Letters of Charles Darwin* 2nd edn 3 vols (London: John Murray)

Daubeny C B 1831 *An Introduction to the Atomic Theory* (London: John Murray) (Royal Institution copy contains bound copy of letter Daubeny to Prout 27 October 1831)

—— 1850 *An Introduction to the Atomic Theory* 2nd edn (Oxford: Oxford University Press)

—— 1852 On the great principles either suggested or worked out by the late celebrated Dr William Prout *Edinb. New Phil. J.* **53** 98–102 (Reprinted in Daubeny (1867) **ii** 123–7)

—— 1867 *Miscellanies* 2 vols (Oxford: James Parker)

Davy Sir H 1800 *Researches, Chemical and Philosophical* (London: J Johnson)

—— 1811 On some new electrochemical researches *Phil. Trans. R. Soc.* 16–74

—— 1816 On the analogies between the undecompounded substances and the constitution of acids *J. Sci. Arts* **1** 288

—— 1858 *Fragmentary Remains Literary and Scientific* (London: John Churchill)

Davy J 1818 On the urinary organs and secretions of some of the amphibia *Phil. Trans. R. Soc.* 303–7

—— 1839–40 *The Collected Works of Sir Humphry Davy Bart.* 9 vols (London: Smith Elder)

Dekosky R K 1973 Spectroscopy and the elements in the late 19th century. The work of Sir William Crookes *Br. J. Hist. Sci.* **6** 400–23

Dempster A J 1918 A new method of positive ray analysis *Phys. Rev.* **11** 316–25

Deville H Sainte-Claire 1857 De la dissociation ou décomposition spontanée des corps sous l'influence de la chaleur *C. R. Acad. Sci., Paris* **45** 857–61

Deville H Sainte-Claire 1860 *Recherches sur la Décomposition des Corps par la Chaleur et la Dissociation* (Paris: C Lahure)
Dingle H 1928 *see* Lockyer with Dingle (1928)
—— 1960 A hundred years of spectroscopy *Br. J. Hist. Sci.* **1** 210
Donovan M 1832 *Treatise on Chemistry* 3rd edn (London: Longman Rees)
Dulong P L 1816 Abstract of Prout (1815–16) *Ann. Chim.* **1** 411–16
Dumas J B 1834 D'analyse eléméntaire des substances organiques *J. Pharm.* **20** 129–56
—— 1842 Recherches sur las composition de l'eau *C. R. Acad. Sci., Paris* **14** 537–47
—— 1837 *Leçons sur la Philosophie Chimiques* (Paris: Bechet)
—— 1859 Mémoire sur les équivalents des corps simple *Ann. Chim.* **55** 129–210
—— 1873 *Letter to Lockyer* 11 September (Sidmouth)
—— 1878 Sur la presence de l'oxygène dans l'argent *C. R. Acad. Sci., Paris* **86** 65–71
Dumas J B and Stas J S 1840 Recherches sur le véritable poids atomique du carbone *C. R. Acad. Sci., Paris* **11** 287–9, 991–1008
Eidinoff M L 1948 The search for tritium *J. Chem. Ed.* **25** 31–4, 40
Einstein A 1905 Zur Elektrodynamik beweigter Körper *Ann. Phys., Lpz.* **17** 891–921
Elliotson J 1828 [Blumenbach's] *Elements of Physiology* 4th edn (London: Longman)
Enys J D 1877 *Correspondence regarding the Appointment of the Writers of the Bridgewater Treatises between Davies Gilbert and Others* (Penryn, Cornwall: J Gill)
Erdmann O L and Marchand R F 1841 Ueber das Atomgewicht das Kohlenstoffes *J. Prakt. Chem.* **23** 159–89 (Engl. Transl. *Phil. Mag.* **18** 332)
—— 1842 Ueber die Atomgewichte des Wasserstoffes *J. Prakt. Chem.* **26** 461–78
Evans Rev J 1816 *History of Bristol* 2 vols (Bristol: W Sheppard)
Eyre J V and Rodd E H 1947 Raphael Meldola (in Findlay and Mills (1947) 98–9)
Fajans K 1913a Die radioaktiven Umwandlungen und die Valenzfrage vom Standpunkte der Struktur der Atome *Verh. Dtsch. Phys. Ges.* **15** 240–59
—— 1913b Über das Uran X_2—das neue Element der Uranreihe *Phys. Z.* **14** 877–84
—— 1923 *Radioactivity* (London: Methuen)
Faraday M 1838 Experimental investigations in electricity *Phil. Trans. R. Soc.* 138
—— 1870 *Life and Letters of Faraday* 2 vols (London: Longman Green)
Farber E 1964 The theory of the elements and nucleosynthesis in the 19th century *Chymia* **9** 181–200
Farrar W V 1965 Nineteenth-century speculations on the complexity of the chemical elements *Br. J. Hist. Sci.* **2** 297–323
Findlay A and Mills W H 1947 *British Chemists* (London: Chemical Society)
Fletcher J 1837 *Rudiments of Physiology* 2nd edn (Edinburgh: J Carfrae)

Fourcroy A F 1801–2 *Systême de Connaissances Chimique* 11 vols (Paris: Baudouin)
Fowler W A 1964 The origin of the elements *Proc. Natl. Acad. Sci. USA* **52** 525–48
Fownes G 1839 The equivalent of carbon *Phil. Mag.* **15** 62–5
Frankland P F and Aston F W 1901 Influence of a heterocyclic group on rotatory power: the ethyl and methyl esters of dipyromucyltartaric acid *Trans. Chem. Soc.* **79** 511–20
Friswell R J 1908 Obituary *Nature* **77** 349
Fyfe A 1814 De Copia Acidi Carbonici e Pulmonibus inter Respirandum Evoluti *MD Dissertation* Edinburgh University (copy given to Prout in Wellcome Institute Library)
Gale G 1974 Chew's monadology *J. Hist. Ideas* **35** 339–48
Gaudin A M 1833 Recherches sur la structure intime des corps inorganiques définis *Ann. Chim.* **52** 113–33
Gay-Lussac J L 1808 Sur la combinaison des substances gaseuses *Mém. Soc. Arcueil* **2** 207–34 (translation 1950 *Foundations of the Molecular Theory* Edinburgh: Alembic Club)
—— 1815 Recherches sur l'acide prussique *Ann. Phil.* **95** 136–231
—— 1816 On uric acid *Ann. Phil.* **7** 350
Gay-Lussac J L and Thenard L G 1811 *Recherches Physico-Chimiques* 2 vols (Paris: Deterville)
Gell-Mann M 1964 A schematic model of baryons and mesons *Phys. Lett.* **8** 214–15 (reprinted in Lichtenberg and Rosen 1980)
Gell-Mann M and Ne'eman Y 1961 *The Eightfold Way* (New York: W A Benjamin)
Giauque W F and Johnston H L 1929 An isotope of oxygen *Nature* **123** 318 (O^{18}), 831 (O^{17})
Gilbert D 1828 Award of Copley medal to Prout *Phil. Mag.* **3** 61–2
Gillispie C C 1951 *Genesis and Geology* (Cambridge, MA: Harvard University Press)
Goldstein E 1886 Ueber eine noch nicht untersuchte Strahlungsform an der Kathode inducirter Entladungen *Berlin Akad. Sber.* **39** 691–9
Graham T 1829 A short account of experimental researches on the diffusion of gases *J. Soc. Arts* **27** 74
—— 1850 *Elements of Chemistry* 2nd edn (London: Ballière)
—— 1863 On the molecular mobility of gases *Proc. R. Soc.* **12** 620–3
Greenaway F 1962 A Victorian scientist: the experimental researches of Sir William Crookes *Proc. R. Inst.* **39** 172–98
Grove R 1976 *Cambridgeshire Coprolites Mining Rush* (Cambridge: Oleander Press)
Gundry D M 1946 The Bridgewater treatises and their authors *History* **31** 143–8
Hahn O 1905 A new radioactive element which evolves thorium emanation *Nature* **71** 574
—— 1907 Ein neues Zwischenprodukt im Thorium *Ber. Dtsch. Chem. Ges.* **40** 1462–8 (Translation Romer (1964) 139–44)
Hall C 1861 *Memoirs of Marshall Hall* (London: R Bentley)

Hanson N R 1963 *The Concept of the Positron* (Cambridge: Cambridge University Press)

Harkins W D 1920 The separation of the element chlorine into normal chlorine and meta-chlorine, and the positive electron *Nature* **105** 230–1

Harkins W D and Wilson E D 1915 The changes of mass and weight involved in the formation of complex atoms *J. Am. Chem. Soc.* **37** 1367–83

Harris J 1775 *Philosophical Arrangements* (London: John Nourse)

Heaviside O 1893–1912 *Electromagnetic Theory* 3 vols (London: Benn)

Heilbron J L 1974 *H. G. J. Moseley. The Life and Letters of an English Physicist* (Berkeley: University of California Press)

Henderson A 1824 *History of Wines Ancient and Modern* (analytical appendix by Prout) (London: Baldwin, Craddock and Joy)

Hendrick R E and Murphy A 1981 Atomism and the illusion of crisis *Phil. Sci.* **48** 454–68

Henry W 1819 On uric acid *Med. Chir. Trans.* **10** 125

—— 1823 *Elements of Experimental Chemistry* 9th edn (London: Baldwin)

—— 1829 *Elements of Experimental Chemistry* 2 vols 11th edn (London: Baldwin and Craddock)

Henry W C 1834 Remarks on the atomic constitution of elastic fluids *Phil. Mag.* **5** 33–9

—— 1837 *Letter to Liebig* 2 June (Staatsbibliothek Munich)

—— 1854 *Memoir of the Life and Scientific Researches of John Dalton* (Cavendish Society: London) 230

Herapath J 1821 A mathematical inquiry into the causes, laws and principal phenomena of heat, gases, gravitation *Ann. Phil.* **17** 273–93

—— 1822 On the physical properties of gases *Ann. Phil.* **19** 16–28

Herschel J W 1824 On certain motions produced in fluid conductors when transmitting the electric current *Phil. Trans. R. Soc.* 162–96

Hodges E 1932 Horton court and its associations *Proc. Clifton Antiquarian Club* **3** 56

Holloway J 1965 *The Victorian Sage* (New York: Norton)

Holloway J 1968 *Noble-Gas Chemistry* (London: Methuen)

Holmes F L 1965 Introduction to Liebig (1842b)

—— 1974 *Claude Bernard and Animal Chemistry* (Cambridge, MA: Harvard University Press)

Homberg E 1983 The influence of demand on the emergence of the dye industry *J. Soc. Dyers Col.* **99** 325–33

Houstoun R A 1923 *Light and Colour* (London: Longman Green)

Huggins W and Miller W A 1864 On the spectra of some fixed stars *Phil. Trans. R. Soc.* **154** 413–36

Hunt T S 1867 Chemistry of the primeval earth *Chem. News* **15** 315–17

—— 1882 Celestial chemistry from the time of Newton *Chem. News* **45** 74–6

—— 1887 *A New Basis for Chemistry; a Chemical Philosophy* (Boston: Cassino)

Hurlbutt R H 1965 *Hume, Newton and the Design Argument* (Lincoln, Nebraska: University of Nebraska Press)

Inkster I and Morrell J B 1983 *Metropolis and Province. Science in British Culture 1780–1850* (London: Hutchinson)

Jacques J 1950 Le vitalisme et la chimie organique pendant la première moitié du XIXe siècle *Rev. Hist. Sci.* **3** 32–66

Jacyna L S 1983 Immanence or transcendence. Theories of life and organization in Britain 1790–1835 *ISIS* **74** 311–29
Jevons W S 1886 Letter of 14 November quoted in *Letters and Journal of W. Stanley Jevons* (London: Macmillan) 393
Johnston J F W 1832 Report on the present state of chemistry *Br. Assoc. Rep.* 414–529
Jones H B 1845 Contributions to the chemistry of the urine *Phil. Trans. R. Soc.* 335–49
—— 1850 *On Animal Chemistry and its Application to Stomach and Renal Diseases* (London: Churchill)
Kasich A H 1946 Prout and the discovery of hydrochloric acid in the gastric juice *Bull. Hist. Med.* **20** 340–58
Kauffman G B 1983a The atomic weight of lead of radioactive sources *J. Chem. Ed.* **59** 3–8
—— 1983b The atomic weight of lead of radioactive sources. *J. Chem. Ed.* **59** 119–23
—— 1983c The mystery of Stephen H. Emmens: successful alchemist or ingenious swindler *Ambix* **30** 65–88
Kayser H 1899 *Letter to Lockyer* 11 November (Sidmouth)
Kendall J 1949 Leo Hendrik Baekeland and the development of phenolic plastics *Chem. Ind.* 67–70
—— 1949–52 Adventures of an hypothesis [Prout's] *Proc. R. Soc. Edinb.* **63A** 1–17
King L S 1963 *The Growth of Medical Thought* (Chicago: Chicago University Press)
Knight D M 1967 *Atoms and Elements* (London: Hutchinson)
—— 1968a Professor Baden Powell and the inductive philosophy *Durham Univ. J.* **60** 81–7
—— 1968b *Classical Scientific Papers: Chemistry* (London: Mills and Boon)
—— 1970 *Classical Scientific Papers Second Series. On the Nature and Arrangement of the Chemical Elements* (London: Mills and Boon)
—— 1978 *The Transcendental Part of Chemistry* (Folkestone: Dawson)
Kopp H 1873 *Die Entwickelung der Chemie* (München: Oldenbourgh)
Kragh H 1982 Julius Thomsen and 19th-century speculations on the complexity of atoms *Ann. Sci.* **39** 37–60
Krätz O P and Priesner C 1983 *Liebigs Experimentalvorlesung* (Weinheim: Verlag Chemie)
Lancet 1850 William Prout 13 April 449
Larmor J 1907 *Memoir and Correspondence of Sir G G Stokes* 2 vols (London: Cambridge University Press)
Lavoisier A 1790 *Elements of Chemistry* (Edinburgh: Creech) (reprint 1965 New York: Dover)
Lederman L M 1978 The upsilon particle *Sci. Am.* **239** 60–8
Lenoir T 1982 *The Strategy of Life: Teleology and Mechanism in 19th Century German Biology* (Dordrecht: Reidel)
Leprince-Ringuet L 1949 Photographic evidence for the existence of a very heavy meson *Rev. Mod. Phys.* **21** 42–3
Leuret F and Lassaigne J L 1825 *Recherches physiologiques et chimiques pour servir à l'Histoire de la Digestion* (Paris: Huzard)

Lichtenberg D B and Rosen S P 1980 *Developments in the Quark Theory of Hadrons. A Reprint Collection* (Nonantum, MA: Hadronic Press)

Liebig J 1831 Über einen neuen Apparat zur Analyse organischer Körper *Pogg. Ann. Phys. Chem.* **31** 1–43

—— 1836 *Letter to C F Mohr* 28 December (in Krätz and Priesner (1983) 14)

—— 1837 *Letter to Berzelius* 26 November (in Tilden (1921) 193)

—— 1842a *Chemistry and its Application to Agriculture and Physiology* 2nd edn (London: Taylor and Walton)

—— 1842b *Animal Chemistry* (reprinted 1965 New York: Johnston Reprint Corp.)

Liebig J and Redtenbacher J 1841 Über das Atomgewicht des Kohlenstoffs *Ann. Chem.* **38** 113–40

Liebig J and Wöhler F 1838 Untersuchungen über die Natur der Harnsäure *Ann. Pharm.* **26** 241–340

Lindemann F A and Aston F W 1919 The possibility of separating isotopes *Phil. Mag.* **37** 523–34

Lipman T O 1964 Wöhler's preparation of urea and the fate of vitalism *J. Chem. Ed.* **41** 452–4

Liveing G D and Dewar J 1881 Identity of the spectral lines of different elements *Proc. R. Soc.* **32** 225–31

—— 1915 *Collected Papers on Spectroscopy* (London: Cambridge University Press)

Lockyer J N 1874a *Contributions to Solar Physics* (Macmillan: London)

—— 1874b Researches on spectrum analysis *Phil. Trans. R. Soc.* **164** 479–94

—— 1878–9 Researches in spectrum analysis *Proc. R. Soc.* **28** 157–80

—— 1879a The spectrum of sodium *Chem. News* **39** 243

—— 1879b Report on dissociation *Br. Assoc. Rep.* 317 (reprint in *Chem. News* **40** 101)

—— 1879c The chemical elements *Nineteenth Century* **5** 285–99

—— 1886 *Studies in Spectrum Analysis* 4th edn (London: Macmillan)

—— 1887 *Chemistry of the Sun* (London: Macmillan)

—— 1900 *Inorganic Evolution* (London: Macmillan)

Lockyer J N and Frankland E 1869 Researches on gaseous spectra in relation to the physical constitution of the sun *Proc. R. Soc.* **17** 288–91 (reprinted in Lockyer (1874a) 525)

Lockyer T M and W L with Dingle H 1928 *Life and Work of Sir Norman Lockyer* (London: Macmillan)

Low D 1848 *Inquiry into the Nature of the Simple Bodies of Chemistry* 2nd edn (London)

Löw R 1980 The progress of organic chemistry during the period of German romantic Naturophilosophie *Ambix* **27** 1–10

Lucretius 56BC *De Rerum Natura* (translated 1951 *On the Nature of Things* (Harmondsworth: Penguin)

McCollum E V 1957 *A History of Nutrition* (Boston: Houghton Mifflin)

McCoy H and Ross W 1907 The specific radioactivity of thorium and the variation of the activity with chemical treatment and with time *J. Am. Chem. Soc.* **29** 1709–18

McGucken W 1969 *Nineteenth-century Spectroscopy* (Baltimore: Johns Hopkins Press)

MacLeod R M 1983 Whigs and Savants: Reflections on the Reform Movement in the Royal Society 1830–48 (In Inkster and Morrell (1983) 55–90)
McMenemey W A 1958 Wilson Philip *J. Hist. Med.* **13** 289–328
Mallet J W 1901 Stas memorial lecture *Memorial Lectures delivered before the Chemical Society 1893–1900* (London: Chemical Society)
Mani N 1956 Das Werk von Friedrich Tiedemann und Leopold Gmelin *Gesnaurus* **13** 190–214
Marcet A 1817 *Essay on Chemical History and Medical Treatment of the Calculus* (London: Longman Hurst)
—— 1823 A singular variety of urine which turned black soon after being discharged *Med. Chin. Trans.* **12** 37–43 (Prout 43–5)
Marchand R F 1844 *Lehrbüch der physiologischen Chemie* (Berlin: Simion)
Marignac J C G 1843 Analyses diverses destinées à la vérification de quelques équivalents chimiques *Bibl. Univ.* **46** 350–77
—— 1860 Commentary on Stas' researches on the mutual relation of atomic weights *Bibl. Univ.* **9** 97–107 (translated in Prout (1932) 48–58)
—— 1902 *Oeuvres Complètes* 2 vols (Geneva: Eggimann)
Marshak R E and Bethe H A 1947 On the two-meson hypothesis *Phys. Rev.* **72** 506
Maumené E J 1846 Sur les équivalents chimiques *Ann. Chim.* **18** 41–79
Maxwell J C 1875 The atom *Encyclopaedia Britannica* 9th edn vol 3 42–3 (reprinted in Niven (1927) **ii** 462)
Mayo H 1833 *Outlines of Physiology* 3rd edn (London: Burgess and Hill)
Meadows A J 1972 *Science and Controversy. A Biography of Sir Norman Lockyer* (London: Macmillan)
Mehra J 1975 *The Solvay Conferences on Physics* (Dordrecht: Reidel)
Meinecke L 1816 Das specifische Gewicht der elastischen Flüssigkeiten *Gilberts Ann. Phys., Lpz.* **24** 159–75
—— 1818 Ueber die Dichtigkeit der elastischflüssigen Körper *Schweiggers J. Chem. Phys.* **22** 137–59
—— 1819 Ueber den stöchiometrischen Werth der Körper *Schweiggers J.* **27** 39–47
Meldrum A 1904 *Avogadro and Dalton* (Aberdeen: The University)
Mellor C M and Cardwell D S L 1963 Dyes and dyeing 1775–1860 *Br. J. Hist. Sci.* **1** 275–9
Mendeleeff D 1905 Faraday lecture in *Principles of Chemistry* 5th edn 2 vols London: Longman Green) **ii** 489–508
Mendelsohn E 1964 *Heat and Life* (Cambridge, MA: Harvard University Press)
—— 1965 Physical models and physiological concepts. Explanation in 19th-century biology *Br. J. Hist. Sci.* **2** 202–19
von Meyer E 1891 *A History of Chemistry* (London: Macmillan)
Miers J 1814 On the composition of azote *Ann. Phil.* **3** 364–72; **4** 180, 260
Millard W F 1955 The life and chemical works of Thomas Thomson *MSc Thesis* University of London
Miller D J 1984 Nearly the 'top' conference *Nature* **311** 210–11
Miller D P 1983 Between hostile camps: Sir Humphry Davy's Presidency of the Royal Society of London 1820–1827 *Br. Soc. Hist. Sci.* **16** 1–47

Mills E J 1873 On the idea of motion *Phil. Mag.* **46** 398–405
Mitchell P H 1956 *Textbook of General Physiology* 5th edn (New York: McGraw-Hill)
Moncrieff R W 1951 *The Chemical Senses* 2nd edn (London: Hill)
Morrell J B 1968–9 Thomas Thomson: Professor of chemistry and university reformer *Br. J. Hist. Sci.* **4** 247
—— 1971 The university of Edinburgh in the late 18th century *ISIS* **62** 158–71
—— 1972 The chemist breeders: the research schools of Liebig and Thomas Thomson *Ambix* **19** 1–46
Morrell J B and Thackray A 1981 *Gentlemen of Science* (London: Clarendon Press)
Morselli M 1984 *Amedeo Avogadro* (Dordrecht: Reidel)
Moseley H G J 1913–14 The high-frequency spectra of the elements *Phil. Mag.* **26** 1024–34; **27** 703–13
Mossotti O 1836 On the forces which regulate the internal constitution of bodies *Taylor's Sci. Mem.* **1** 448–69
Müller J 1838 *Handbüch der Physiologie des Menschen* 3rd edn (Coblenz: Hölscher) [translated *Elements of Physiology* 2 vols (London: Taylor and Walton)]
Munk W 1878 William Prout in *Roll of the Royal College of Physicians* 3 vols (London: The Royal College of Physicians) **iii** 109–13
Nambu Y 1976 The confinement of quarks *Sci. Am.* **235** 48–60
Needham J and Pagel W 1940 *Background to Modern Science* 2nd edn (London: Cambridge University Press)
Newton I 1744 *Opuscula* ed Joh Castillioneus **ii** *Lectionis Opticae* (Lausanne: Marcum-Michaelem Bousequet)
—— 1952 *Opticks* reprint of 1704 edition (New York: Dover)
—— 1959 *Correspondence of Isaac Newton* vol 1 (London: Cambridge University Press)
Nier A O 1937 A mass-spectrographic study of the isotopes *Phys. Rev.* **52** 933–7
Niven W D 1890 *Scientific Papers of James Clerk Maxwell* 2 vols (London: Cambridge University Press)
Nobel Lectures in Chemistry 1966 2 vols (Amsterdam: Elsevier)
Oersted H C 1820 Experiments on the electrical conflict in the neighbourhood of a magnet *Ann. Phil.* **16** 273–6
Oesper R E and LeMay P 1950 Henri Sainte-Claire Deville 1818–1881 *Chymia* **3** 220
Pahl M and Weimer U 1958 Mass spectrometric identification of $(NeH)^+$ *Z. Naturf.* **13A** 745
Panton D R 1981 Mass-spectrometry. The Development of a Scientific Community *MPhil Thesis* University of Surrey 2 vols
Panton D R and Reuben B G 1973 The sociological interpretation of scientific progress: the case of mass spectrometry *Adv. Mass Spectrom.* **6** 975–9
Partington J R 1962 *A History of Chemistry* vol 3 (London: Macmillan)
—— 1964 *A History of Chemistry* vol 4 (London: Macmillan)

Pelling M 1978 *Cholera, Fever and English Medicine 1825–1865* (Oxford: Oxford University Press)
Pelouze T J 1845 Mémoire sur les équivalents de plusieurs corps simples *C. R. Acad. Sci., Paris* **20** 1047–55
Penny F 1839 Determination of several equivalent numbers *Phil. Trans. R. Soc.* 13–33
Philip A P W 1831 Dispute with Prout *Med. Gaz.* **8** 641–52, 737–40, 770–6, 801–2, 843–4; **9** 69–79
Phillips R 1824 Table of equivalent weights *Ann. Phil.* **23** 185–97
—— 1825a Table of Berzelius' equivalents *Ann. Phil.* **25** 439–54
—— 1825b Review of Thomson's *First Principles, Ann. Phil.* **26** 147
—— 1826 Silliman's review of Thomson's results *Ann. Phil.* **27** 68
—— 1828 Attack of Berzelius on Dr Thomson *Phil. Mag.* **4** 450–3
—— 1839 Researches on chemical equivalents *Phil. Trans. R. Soc.* 35–8
—— 1844 Review of Low *Phil. Mag.* **24** 296
Pickering A 1984 *Constructing Quarks. A Sociological History of Particle Physics* (Edinburgh: Edinburgh University Press)
Polkinghorne J C 1979 *The Particle Play* (Oxford: W H Freeman)
Prideaux J 1830 On the composition of barium chloride *Phil. Mag.* **7** 276–80
Prout W 1810a *De Facultate Sentiendi* (unpublished MS dated 'Edinburgh, 1810' Wellcome Institute Library)
—— 1810b *Letter to Walter Adam* 21 August (Ballogie papers)
—— 1811a *Dissertatio Medica Inauguralis De Febribus Intermittentibus* (Edinburgh)
—— 1811b *Letter to Mrs Adam* 12 September (Ballogie papers)
—— 1812a Observations upon the sensations of taste and smell *London Med. Phys. J.* **28** 457–61
—— 1812b *Letter to Mrs Adam* 24 April (Ballogie papers)
—— 1813a Observations on the quantity of carbonic acid gas emitted from the lungs during respiration at different times and under different circumstances *Ann. Phil.* **4** 331–7
—— 1813b *Letter to Walter Adam* 19 September (Ballogie papers)
—— 1813c On injecting the blood vessels *London Med. Phys. J.* **30** 89–96
—— 1814a Further remarks on the quantity of carbonic acid gas emitted from the lungs during respiration *Ann. Phil.* **4** 331–7
—— 1814b *Letter to Mrs Adam* 22 April (Ballogie papers)
—— 1814c *Letter to Walter Adam* 23 April (Ballogie papers)
—— 1814d MSS of private lectures on animal chemistry (Wellcome Institute Library)
—— 1815a On the analysis of organic substances *Ann. Phil.* **6** 269–73
—— 1815b Analysis of the excrement of the boa constrictor *Ann. Phil.* **5** 413–16
—— 1815c On the relations between the specific gravities of bodies and the weights of their atoms *Ann. Phil.* **6** 321–30, 472 (reprinted in Prout (1932))
—— 1815–40 Abstract of Prout (1827) *Proc. R. Soc.* **2** 324–6
—— 1816a Inquiry into the origins and properties of blood *Ann. Med. Surg.* **1** 10–26, 133–57, 277–89 (republished as Prout (1819a))

Prout W 1816b Correction of a mistake in the essay on the relation between the specific gravities of bodies *Ann. Phil.* **7** 111–13 (reprinted in Prout (1932))

—— 1817 Observations on the nature and proximate principles of the urine with a few remarks upon the means of preventing those diseases connected with a morbid state of that fluid *Med. Chir. Trans.* **8** 526–49

—— 1818a Further observations on the proximate principles of the urine *Med. Chir. Trans.* **9** 472–84

—— 1818b Description of an acid principle prepared from lithic or uric acid *Phil. Trans. R. Soc.* 420–8

—— 1819a On sanguification *Ann. Phil.* **13** 12–25, 265–79

—— 1819b Description of an urinary calculus composed of the lithate or urate of ammonia *Med. Chir. Trans.* **10** 389–95

—— 1820a Excrement of the chamaelonis vulgaris *Ann. Phil.* **15** 471

—— 1820b Description of an apparatus for the analysis of organized substances *Ann. Phil.* **15** 190–2

—— 1820c Tests of pure potash *Ann. Phil.* **16** 150

—— 1821 *An Inquiry into the Nature and Treatment of Gravel, Calculus and other Affections of the Urinary Organs* (London: Baldwin, Craddock and Joy)

—— 1822a On the ultimate analysis of animal and vegetable substances *Ann. Phil.* **20** 424–5

—— 1822b Observations on the changes the egg undergoes during incubation in the common fowl *Phil. Trans. R. Soc.* 377–400

—— 1824a On the nature of the acid and saline matters usually existing in the stomachs of animals *Phil. Trans. R. Soc.* 45–9

—— 1824b *Letter to Walter Adam* 15 July (Ballogie papers)

—— 1825a *An Inquiry into the Nature and Treatment of Diabetes, Calculus and other Affections of the Urinary Organs* 2nd edn (London: Baldwin, Craddock and Joy)

—— 1825b Description of an instrument for ascertaining the specific gravity of the urine in diabetes *Ann. Phil.* **25** 334–5

—— 1826 Remarks on certain observations made by MM Leuret and Lassaigne, and Professors Tidemann and Gmelin, in their works on digestion *Ann. Phil.* **26** 405–10

—— 1827 On the ultimate composition of simple alimentary substances with some remarks on the analysis of organized bodies in general *Phil. Trans. R. Soc.* 355–88

—— 1828 Further remarks on Messrs Tiedemann and Gmelin's observations of stomach acids *Phil. Mag.* **4** 120–3

—— 1830 *Letter to Davies Gilbert* 8 October (Royal Inst. Cornwall)

—— 1831a Gulstonian Lectures: observations on the application of chemistry to physiology, pathology and practice *Med. Gaz.* **8** 257–65, 321–7, 385–91

—— 1831b *Letter to Daubeny* 12 September in Daubeny (1831)

—— 1831c Reply to Philip *Med. Gaz.* **8** 705–7, 769–70, 802–4; **9** 38–46

—— 1834a *Chemistry, Meteorology and the Function of Digestion considered with reference to Natural Theology* (London: W Pickering)

—— 1834b Reply to Dr Henry *Phil. Mag.* **5** 132–3

—— 1840 *On the Nature and Treatment of Stomach and Urinary Diseases* 3rd edn (London: J Churchill)
—— 1843 *On the Nature and Treatment of Stomach and Urinary Diseases* 4th edn (London: J Churchill)
—— 1845 *Chemistry, Meteorology and the Function of Digestion* 3rd edn (London: J Churchill)
—— 1848 *On the Nature and Treatment of Stomach and Renal Diseases* 5th edn (London: J Churchill)
—— 1855 *Bridgewater Treatise: Chemistry, Meteorology and the Function of Digestion* 4th posthumous edn (London: Bohn Library)
—— 1932 *Prout's Hypothesis* (Edinburgh: Alembic Club)
Rainy H 1825 On the specific gravity of hydrogen gas *Ann. Phil.* **26** 135–7
—— 1826 Further remarks on the specific gravity of hydrogen gas and on Prout's modification of the atomic theory *Ann. Phil.* **27** 187–94
Ramsey W and Soddy F 1903 Experiments in radioactivity and the production of helium from radium *Chem. News* **88** 100–1
Reines F and Cowan C L 1956 Detection of the free neutrino—a confirmation *Science* **124** 103–4
Reynolds J E 1886 Note on a method of illustrating the periodic law *Chem. News* **54** 1–4
Rigg R 1844 *Experimental Researches shewing Carbon to be a Compound Body made by Plants* (London: Smith Elder)
Roberton J 1831 On the organic agent of Dr Prout *Med. Gaz.* **8** 744–8
—— 1836 *Critical Remarks on Certain Published Opinions concerning Life and Mind* (London: Longman Rees)
Roberts G K 1976 The establishment of the Royal College of Chemistry *Hist. Stud. Phys. Sci.* **7** 437–86
Rocke A J 1978 Atoms and equivalents. The early development of the chemical atomic theory *Hist. Stud. Phys. Sci.* **9** 225–63
—— 1984 *Chemical Atomism in the Nineteenth Century: From Dalton to Cannizzaro* (Columbus, Ohio: Ohio State University Press)
Romer A 1964 *The Discovery of Radioactivity and Transmutation* (New York: Dover)
Roscoe H E 1873 *Letter to Lockyer* 7 December (Sidmouth)
—— 1874 *Letter to Lockyer* 26 May (Sidmouth)
—— 1878a *Letter to Lockyer* 25 March (Sidmouth)
—— 1878b *Letter to Lockyer* 5 December (Sidmouth)
—— 1878c *Letter to Lockyer* 18 December (Sidmouth)
—— 1879 *Letter to Lockyer* 10 February (Sidmouth)
—— 1880 *Letter to Lockyer* 11 April (Sidmouth)
Rowe J S 1955 Chemical Studies at University College London *PhD Thesis* University of London
Russell C A 1963 The electrochemical theory of Berzelius *Ann. Sci.* **19** 117–26, 128–45
—— 1968 Berzelius and the development of the atomic theory (in Cardwell (1968) 259–73)
Rutherford E 1899 Uranium radiation and the electrical conduction produced by it *Phil. Mag.* **47** 392–407; *Collected Papers* **i** 105–18

Rutherford E 1900 A radioactive substance emitted from thorium compounds *Phil. Mag.* **49** 1–14; *Collected Papers* **i** 220–31
—— 1903 The magnetic and electric deviation of the easily absorbed rays from radium *Phil. Mag.* **5** 177–87; *Collected Papers* **i** 549–57
—— 1911 The scattering of the α and β Rays by matter and the structure of the atom *Phil. Mag.* **21** 669–88; *Collected Papers* **ii** 238–54
—— 1914 The structure of the atom *Phil. Mag.* **27** 488–98
—— 1919 Collision of particles with light atoms *Phil. Mag.* **37** 537–87; *Collected Papers* **ii** 547–90
—— 1920a Physics at the British Association *Nature* **106** 357
—— 1920b The nuclear constitution of atoms *Proc. R. Soc.* A **97** 374–400
—— 1937 The search for the isotopes of hydrogen and helium of mass 3 *Nature* **140** 303–5
—— 1940 Forty years of physics (in Needham and Pagel (1940) 49–76)
—— 1962–5 *The Collected Papers of Lord Rutherford of Nelson* 3 vols (New York: Interscience)
Rutherford E and Geiger H 1908 The charge and nature of the α-particle *Proc. R. Soc.* A **81** 162–73
Rutherford E and Soddy F 1902 The radioactivity of thorium compounds. The cause and nature of radioactivity *Trans. Chem. Soc.* **81** 837–60 (reprinted in Romer (1964) 87–116)
Sakata S and Inowe T 1946 The two-meson hypothesis *Prog. Theor. Phys.* **1** 143
Scholes P 1970 *The Oxford Companion to Music* 10th edn (London: Oxford University Press)
Schrader-Frechette K 1977 Atomism in crisis: an analysis of the current high energy paradigm *Phil. Sci.* **44** 409–40
Schwann T 1836 Ueber das Wesen des Verdauungsprocesses *Arch. Anat. Phys.* 90–138
Scudamore C 1817 *A Treatise on the Nature and Cure of Gout* 2nd edn (London: Longman Hurst)
Siegel D M 1978 Classical electromagnetic and relativistic approaches to the problem of nonintegral atomic mass *Hist. Stud. Phys. Sci.* **9** 323–60
Siegfried R 1956 The chemical basis of Prout's hypothesis *J. Chem. Ed.* **33** 263–6
—— 1959 The chemical philosophy of Davy *Chymia* **5** 193–6
—— 1963 The discovery of potassium and sodium and the problem of the chemical elements *ISIS* **54** 247–58
——1964 The phlogistic conjectures of Humphry Davy *Chymia* **9** 117–24
Singer C 1921 *Studies in History and Method of Science* 2 vols (Oxford: Clarendon Press)
Smith E 1859 Experimental inquiries into the chemical and other phenomena of respiration *Phil. Trans. R. Soc.* 681–714, 715–42
Snyder J N 1950 On the changing status of mesons *Am. J. Phys.* **18** 41–9
Soddy F 1911a Radioactivity *Ann. Rep. Prog. Chem.* (for 1910) **8**
—— 1911b The chemistry of mesothorium *J. Chem. Soc.* **99** 72–83 (reprinted in Romer (1964) 179–90)
—— 1911c *The Chemistry of the Radio-elements* (London: Longman Green)

—— 1913a The intra-atomic charge *Nature* **92** 400
—— 1913b Radioactivity *Ann. Rep. Prog. Chem.* (for 1912) **10**
—— 1932 *The Interpretation of the Atom* (London: John Murray)
—— 1966 Nobel speech (1921): The origins of the concept of isotopes (in *Nobel Lectures in Chemistry* 1966 **i** 371–99)
Somerville Mrs M 1842 *On the Connexion of the Physical Sciences* 6th edn (London: John Murray)
Spencer H 1891 *Essays: Scientific, Political, Speculative* 3 vols (London: Williams and Norgate)
van Spronsen J W 1969 *The Periodic Systems of Chemical Elements. A History of the First Hundred Years* (Amsterdam: Elsevier)
Starling E and Evans C L 1968 *Principles of Human Physiology* 14th edn (London: Churchill)
Stas J S 1860 Researches on the mutual relations of atomic weights *Bull. Acad. R. Belg.* **10** 208–336 (translated in Prout (1932) 41–7)
Stewart B and Tait P G 1875 *The Unseen Universe* (London: Macmillan)
Stokes G 1879 *Letter to Lockyer* 27 January (in Larmor (1907) **i** 406)
Stoney G J 1894 Of the 'electron' or atom of electricity *Phil. Mag.* **38** 418–20
Strömholm D and Svedberg T 1909 Untersuchungen über die Chemie der radioaktiven Grundstoffe *Z. Anorg. Chem.* **61** 338–46; **63** 197–206
Strutt R J 1901 On the tendency of the atomic weights to approximate to whole numbers *Phil. Mag.* **1** 311–14
Szabadváry F 1966 *History of Analytical Chemistry* (Oxford: Pergamon)
Taylor A 1849 On the alleged production of phosphate of lime and iron in the egg during incubation *Guy's Hosp. Med. Rep.* **6** 141–8
Terrey H 1937 Edward Turner *Ann. Sci.* **2** 137
Thomson G P 1946 Obituary of Francis Aston *Nature* **157** 290–2
Thomson J J 1897 Cathode rays *Phil. Mag.* **44** 293–316
—— 1907 On rays of positive electricity *Phil. Mag.* **13** 561–75
—— 1912 Further experiments on positive rays *Phil. Mag.* **24** 209–53
—— 1913a Some further applications of the method of positive rays *Proc. R. Inst.* **20** 591–600
—— 1913b *Rays of Positive Electricity and their Application to Chemical Analysis* (London: Longman Green)
—— 1914 Rays of positive electricity *Proc. R. Soc.* A **89** 1–20
—— 1918 Problems in atomic structure *Engineering* 29 March 345
Thomson J J and Rutherford E 1896 On the passage of electricity through gases exposed to Röntgen rays *Phil. Mag.* **42** 392–407
Thomson J J *et al* 1921 Discussion on isotopes *Proc. R. Soc.* A **99** 87–94 (Thomson), 95–7 (Aston), 97–100 (Soddy)
Thomson R D 1839 On the proofs of the existence of free muriatic acid in the stomach *Br. Assoc. Rep.* 58
Thomson T 1802 *System of Chemistry* 2 vols (Edinburgh: J Brown)
—— 1813 On the Daltonian theory of definite proportions in chemical combinations *Ann. Phil.* **2** 32–52, 109–15, 167–71, 293–301
—— 1814 On the Daltonian theory *Ann. Phil.* **3** 134–40, 375–8; **4** 11–18. 83–9
—— 1815 Annual report on the progress of chemistry *Ann. Phil* **5** 1–153

Thomson T 1816a Annual report on the progress of chemistry *Ann. Phil.* **7** 1–71
—— 1816b Observations of the relation between specific gravities and atomic weights *Ann. Phil.* **7** 343–6
—— 1817 *System of Chemistry* 5th edn 4 vols (London: Baldwin, Craddock and Joy)
—— 1818 Observations of the weights of the atoms of chemical bodies *Ann. Phil.* **12** 338–50, 436–41
—— 1819 On the specific gravity of hydrogen *Ann. Phil.* **14** 65–6
—— 1820a Historical sketch of progress of chemistry during 1819 *Ann. Phil.* **16** 1–46
—— 1820b Experiments to determine the weight of the atoms of barytes *Ann. Phil.* **16** 323–3
—— 1820c On the specific gravity of gases *Ann. Phil.* **16** 161–77, 241–68
—— 1821a Experiments to determine the true weight of the atoms *Ann. Phil.* **17** 241–52
—— 1821b Experiments to determine the atomic weights of various metals *Ann. Phil.* **18** 120–46
—— 1822a Reply to Ure *Ann. Phil.* **19** 241–75
—— 1822b The influence of humidity in modifying the specific gravity of gases *Ann. Phil.* **19** 302–8
—— 1825a *An Attempt to Establish the First Principles of Chemistry by Experiment* 2 vols (London: Baldwin, Craddock and Joy)
—— 1825b Reply to Rainy *Ann. Phil.* **26** 352–60
—— 1826 Reply to Ure *Ann. Phil.* **27** 1–14
—— 1829 Reply to Berzelius *Phil. Mag.* **5** 217–23
—— 1830–1 *History of Chemistry* 2 vols (London: Colburn and Bentley)
—— 1831 *A System of Chemistry of Inorganic Bodies* 2 vols (London: Baldwin and Craddock)
—— 1836 On atomic weights *Rec. Gen. Sci.* **3** 179–91, 251–3
Tiedemann F and Gmelin L 1827 *Recherches Expérimentales Physiologiques et Chimiques sur la Digestion* 2 vols (Paris: Ballière)
—— 1828 Reply to the remarks of Dr Prout *Phil. Mag.* **4** 3–5
Tilden W A 1921 *Famous Chemists* (London: Routledge)
Todd A C 1967 *Beyond the Blaze. A Biography of Davies Gilbert* (Truro: Barton)
Trefil J S 1980 *From Atoms to Quarks* (London: Athlone Press)
Trenn T J 1974 The justification of transmutation: speculations of Ramsay and experiments of Rutherford *Ambix* **21** 16–26
Turner E 1825 *An Introduction to the Study of the Laws of Chemical Combination and the Atomic Theory* (Edinburgh: Maclachan and Stewart)
—— 1828 Chemical examination of the oxides of manganese *Phil. Mag.* **4** 22–35
—— 1829 On the composition of chloride of barium *Phil. Trans. R. Soc.* 291–300
—— 1832 On atomic weights *Phil. Mag.* **1** 109–12
—— 1833 Experimental researches on atomic weights *Phil. Trans. R. Soc.* 523–44
—— 1834a *Elements of Chemistry* 4th edn (London: J Taylor)

—— 1834b Researches on atomic weights *Phil. Mag.* **4** 397
—— 1835 *Letter to Prout* on chemical symbolism, together with draft reply by Prout 30 January (Ballogie papers)
Turner F M 1981 John Tyndall and Victorian Scientific Naturalism (in Brock, McMillan and Mollan (1981) 169–80)
Tyndall J 1854 *Private Journal* 19 November (Royal Inst. Archives)
Ure A 1821a On equivalents *Phil. Mag.* **57** 95–116
—— 1821b Review of Thomson's *System Q. J. Sci.* **11** 119–71
—— 1822 Ultimate analysis of vegetable and animal substances *Phil. Trans. R. Soc.* 457
—— 1825 Review of Thomson's *First Principles, Q. J. Sci* **20** 113–41
Urey H C, Brickwedde F G and Murphy G M 1932 A hydrogen isotope of mass 2 *Phys. Rev.* **39** 164–5; **40** 1–18
Villard P 1900 Sur la réflexion et la réfraction des rayons cathodique et des rayons déviables du radium *C. R. Acad. Sci., Paris* **130** 1010–12
Wakley T 1844 On the labours of Dr Prout *The Lancet* 6 January 486–90
Weeks E M 1968 *Discovery of the Elements* 7th edn (Easton, PA: Journal of Chemical Education)
Weisskopf V F 1970 Three steps in the structure of matter *Phys. Today* **23** 17–24
Whewell W 1847 *Philosophy of the Inductive Sciences* 2nd edn 2 vols (London: J W Parker) **ii** 413
Whyte L L 1950–1 Fundamental physical theory. An interpretation of the present position of the theory of particles *Br. J. Phil. Sci.* **1** 303–27
Wien W C 1898 Untersuchungen über die elektrische Entladung in verdünnten Gasen *Ann. Phys., Lpz.* **301** 440–52
—— 1902 Ueber die Natur der positiven Elektronen *Ann. Phys., Lpz.* **314** 660–4
Williams L P 1965 *Michael Faraday* (London: Chapman and Hall)
Wilson F L 1968 Fermi's theory of beta decay *Am. J. Phys.* **36** 1150–60
Wilson G 1862 *Religio Chemici* (London: Macmillan)
Witner E E 1946 Integral and rational numbers in the nuclear domain *Proc. Natl. Acad. Sci. USA* **32** 283–8
Wöhler F 1828 On the artificial production of urea *Pogg. Ann.* **12** 253–6
Wollaston W H 1810 On cystic oxide *Phil. Trans. R. Soc.* 223–30
—— 1814 A synoptic scale of chemical equivalents *Phil. Trans. R. Soc.* 1–22
Wollheim R 1963 *Hume on Religion* (London: Collins Fontana)
Wurtz C A 1865 Equivalents, atomic weights and molecular weights *Chem. News* **11** 265–6
Young C A 1872–3 On the bright lines in the spectrum of the solar atmosphere *Nature* **7** 17–20
—— 1879 *Letter to Lockyer* 24 November (Sidmouth)
—— 1882 *The Sun* (London: King) pp 89–101
Yukawa H 1935 Theory of the β-disintegration *Proc. Phys. Math. Soc. Jpn.* **17** 48
Yukawa H and Kikuchi C 1950 The birth of the meson theory *Am. J. Sci.* **18** 154–6
Zweig G 1964 An SU3 model for strong interaction symmetry (in Lichtenberg and Rosen (1980) 22–101)

Index

Abney, Sir William, 192
Adam, Agnes (Mrs Prout), 4, 8–9
Adam, Alexander, 3
Adam, Mrs Alexander, 4–6
Adam, Walter, 4–7, 34, 55
Adrian, Lord E D, 208
affinity, 152, 175
air, composition of, 98
albuminous substances, 57, 122, 127, 132, 136
alchemy, 189, 198
Allen, William, 35, 126
Alvarez, L W, 214
amniotic fluid, 21
Ampère, André Marie, 112, 116–17, 119, 166
analysis, organic, 12–21, 32
Anderson, Carl David, 219
Apjohn, James, 154
Arago, Dominique F J, 100–1, 146
Aristotelian philosophy, vii, 37–8, 83
Armstrong, Henry Edward, 180, 190
Aston, Francis, viii, 182, 217; ancestry and birth, 201–2; works on conductivity, 202–4; discovers dark space, 203–4; works on positive rays, 205–7; isotopes of neon, 206–8; invents mass spectrograph, 208–9; adopts packing concept, 211, 213–14; Nobel prize, 211; proprietorial attitude, 212; misses isotopes of hydrogen and oxygen, 214; cites Prout, 210, 218
atomic number, 201, 212, 217
atomic weights, 98–9, 102–4, 143–4, 149, 162–4, 167–8, 170, 172, 174, 177, 191–2, 201, 217; chlorine, 21, 177–8; oxygen scale, 93; carbon, 139; barium, 160, 162–5; manganese, 162; silver, 178; as averages, 196–7, 200; neon, 206–8, 210; international committee for, 215
Avogadro, Amedeo, 98, 102, 105, 116–19, 121, 166, 172

Badash, Lawrence, 199
Baekeland, Leo, 174
Baillie, Mathew, 5, 8
Bainbridge, K T, 216
Banks, Sir Joseph, 7
barium chloride, 159, 161–3
Barkla, Charles, G, 218
Beaumont, William, 52, 56
Bell, Sir Charles, 62–3
Bernard, Claude, 13, 56, 80, 126
Berthelot, Marcellin, 79, 184–5
Berzelius, Jöns, 6–7, 48–9, 51–3, 58, 83, 93, 97, 101, 106–7, 110, 118–19, 121, 148, 152–3, 161, 166–9, 173; describes Prout, 11; analyses of organic compounds, 14–16; criticises Prout's analyses, 23–5; criticises Thomson, 146–7, 158–

Berzelius, Jöns (contd) 60, 162; his atomic weights examined by Turner, 163–4; on nature of elements, 175–7
Bethe, Hans A, 220
Bichat, Xavier, 71, 73
Biot, Jean Baptiste, 100–1, 146
Bird, Golding, 134
Blane, Sir Gilbert, 8, 50
Bleakney, Walter, 214
Blomfield, Bishop Charles J, 62–3, 80
blood, 7, 47–9, 76, 125
Blumenbach, Johann F, 47, 136
Bohr, Niels, 218
Boltwood, Bertram, 198–9
Boscovich, R G, 150
Boyle, Robert, 84
Brande, William Thomas, 28, 134, 144–5, 148, 152
Brayley, Edward William, 21
Brewster, Sir David, 63
Bridgewater treatises, 61–4
Bright, Richard, 28
British Association for the Advancement of Science, 9, 56, 164–6, 180, 191, 208, 217
Brodie, Sir Benjamin Collins, 10, 134, 180, 182, 191, 195
van den Broek, A, 217
Brown, Adrian J, 202
Brown, Samuel, 175
Buckland, William, 22, 62–3, 68

calcium, transmutation of, 50
calculus, urinary, 27–8, 31, 129–30
Cannizzaro, Stanislao, 131
carbohydrates, 14, 57–8, 120, 122, 129, 132, 136
Carnelley, Thomas, 179, 193–5
Chadwick, Sir James, ix, 219
Chalmers, Thomas, 62–3
Chambers, Robert, 184
chemical change, vii, 82–3, 114, 218
Chemical Club, 7
Chemical Society of London, 9, 134, 180
Chenevix, Richard, 82, 89

Chew, Geoffrey, 223–5
Children, John G, 55, 158
cholera, 10, 129–30
Clark, Thomas, 166
Clarke, Frank W, 182
Cockcroft, Sir John D, 216
Coindet, Jean, 67
completion, 26, 124
conductivity of gases, 202–4
continuity, principle of, 183–6
Cooper, Sir Astley, 6–7, 48
coprolites, 22
Cornug, R, 214
corpuscular philosophy, vii, 82–3
cosmic rays, 220
Crookes, Sir William, 179, 182, 184, 190–2, 194, 198; as sage, 180–1; sees Lockyer as alchemist, 189; supports dissociation, 193; views on elements, 195–7, 208; identifies protyle with electron, 195; dark space, 202–3
Curie, Marie, 197
Curie, Pierre, 197
cuttlefish ink, 21–2

Dalton, John, viii, 11, 89, 116, 133, 148, 152, 155, 171, 205; atomic theory, 23, 83, 96–9, 112, 131, 144; Prout criticises, 118–21; nomenclature, 138–9; specific gravities, 165
Daniel, John Frederick, 177, 179
Darwin, Charles, 61
Daubeny, Charles G B, 21, 31, 50, 58, 118, 120–1, 165, 169–70
Davy, Sir Humphry, 3–4, 8, 11, 27, 85, 93, 96, 100–1, 110–12, 133, 139, 144, 147–8, 152, 158, 175, 191; on analogy, 46; on elements, 72, 88–91, 104–5
Davy, John, 4, 22
Democritus, 83–4
Dempster, A J, 216
deuterium, 214
Deville, H E Sainte-Claire, 182
Dewar, Sir James, 188, 192
diabetes, 25, 129–30

248   Index

digestion, 7, 13, 22, 47–59, 70, 124–6, 135
Dingle, Herbert, 186
dissociation, 179, 185–94
Donovan, Michael, 117
Dulong, Pierre Louis, 100–1, 104, 159
Dumas, Jean Baptiste, 73, 107, 116–17, 125–6, 130, 152, 166, 168, 170, 172; on atomic weight of carbon, 172–3; supports Prout's views, 174; criticises Stas's atomic weights, 191–2; use of hydrocarbon analogy, 193

Edinburgh, University of, 3–4, 11, 34, 36
Egerton, Rev. Francis Henry, Earl of Bridgewater, 61, 63–4
eightfold way, 222–3
Einstein, Albert, 211
electron, 183, 195, 198, 203–5, 218–20, 224; as protyle, 195
elements, viii, 50–1, 72–3, 82, 84–8, 91–2, 94–5, 102–3, 105–7, 149–50, 170, 172; genesis of, 179, 184, 195–7; increase in number, 179; rare earths, 180, 197; dissociation, 185–90, 192; spectra of, 187–9; views of Crookes on 195–7; radio-elements, 198–201; rare (inert) gases, 206; neon, 206–8; *see also* calcium, hydrogen, nitrogen
Elliotson, John, 4, 25, 47, 50–1, 57, 67, 136
Emmens, Stephen, 194
Empedocles, 82, 84
Erdmann, Otto Linné, 173–4
Everett, E, 204–5
evolution, *see* elements, genesis
excrement, of snake, 22, 26

Fajans, Kasimir, 200–1
Faraday, Michael, 95, 110, 134, 172, 202
Fermi, Enrico, 219, 222
Festing, E R, 192
flavour, 46–7

Fletcher, John, 79
de Fourcroy, Antoine François, 13, 73, 126
Fowler, Alfred, 188
Fowler, Sir Ralph Howard, 209
Fowler, W A, 218
Fownes, George, 134, 173
Frankland, Sir Edward, 185, 187, 195
Frankland, Percy F, 202
Friswell, Richard J, 185–6

Gardiner, John H, 191
gastric juice, 48, 52–6, 59, 124, 133
Gaudin, Marc Antoine, 117
Gay-Lussac, Joseph L, viii, 14, 16, 23–5, 97, 100, 103, 116, 148, 171
Gell-Mann, Murray, 222–5
Giauque, William F, 214
Gilbert, Davies, 18, 61–3
Glashow, Sheldon, L, 225
Gmelin, Leopold, 49, 52–6
Graham, Thomas, 134, 178, 207
Greenberg, Oscar, 224
Griffith, John, 56
Guy's Hospital, 5–6

Hahn, Otto, 198
Hall, Marshall, 4, 61
Harkins, William D, 211
Harris, James, 37–8, 85–6
Hayes, John, 8, 10
hearing, 41, 43–4
Heaviside, Oliver, 217
Henderson, Alexander, 52
Henry, William, 16, 28, 145
Henry, William Charles, 11, 19, 117–18, 135
Herapath, John, 172
Herschel, Sir John F W, 58, 62, 77, 146
Holland, Sir Henry, 4
Hope, Thomas Charles, 4, 96
Horton, Gloucestershire, 1, 5, 8
Howley, Archbishop William, 62–3, 80
Huggins, Sir William, 182, 186, 192
Hume, David, 60–1, 66, 70

Hunt, Thomas Sterry, 180, 182, 184
hydrogen, 151, 169, 174, 215, 217; found in elements, 89, 91, 107, 171, 189, 191–2; transmutation into helium, 219; as protyle, 105; specific gravity of, 145, 154–7; isotopes of, 210

injection, anatomical, 35
Inowe, T, 220
iodine, use in goitre, 67
ionisation, 182, 195, 219
isomerism, 58, 121, 170, 172
isomorphism, 120–1, 170, 172, 174
isotopes, 179, 199, 202, 207–15; of oxygen and hydrogen, 214

Jevons, Stanley, 192
Johnston, H L, 214
Johnston, James F W, 117, 166–7
Jones, Henry Bence, 76, 131, 141
Jones, Rev. Thomas, 3, 5

Kayser, Ernst, 182
Kidd, John, 62–3
Kirby, William, 63
Kirchhoff, Gustav R, 184

Lassaigne, J L, 52–4
Lavoisier, Antoine, viii, 13–14, 17, 83–6, 88–9, 107, 126
Lawes, Sir John Bennet, 22
Leuret, F, 52–5
von Liebig, Justus, 13, 27, 32, 73, 75–6, 79, 81, 122, 125–7, 130; on organic analysis, 18–20; invited to England, 132; relations with Prout, 133–41; on carbon, 173; *Multiplenfieber*, 173
Lindemann, Frederick, 208
Liveing, George, 188, 192
Lockyer, Sir Norman J, 179, 182, 195, 198; comparison with Crookes, 180–2; on uniformity of nature, 183–6; dissociation hypothesis, 185–94; effects of impurities, 190–2; uses molecular analogy, 193

Low, David, 175
Lucretius, 221

McCoy, H N, 199
Marcet, Alexander, 6–8, 15, 22–4, 27–8, 31, 48, 52, 73
Marchard, Richard Felix, 136, 173–4
Marignac, J C G, 170, 173; on unit less than hydrogen, 177–8, 210–11
Marshak, R E, 220
mass-spectrometer, 208–9, 211–13, 214–16
Matthey, George, 188
Matthiessen, Augustus, 188
Maumené, E J, 173, 178
Mayo, Herbert, 51
Medical and Chirurgical Society of London, 7, 16, 23, 27, 31
Meinecke, Ludwig, 170–1
melanic acid, 27
Meldola, Raphael, 188–9
Mendeleeff, Dmitri, 104, 150, 184–5, 196, 199, 222–3
merorganisation, 58, 77, 132
mesons, 220, 222
meteorology, 64, 68–9
von Meyer, Ernst, 104
Meyer, Lothar, 150
Meyer, Victor, 191
Miers, John, 93–6, 104, 152, 175
Miller, William Allen, 132, 182
Millikan, Robert, 224
Mills, Edward J, 195
minima naturalis, vii
Monro, Alexander *tertius*, 4
Moseley, Henry G J, 201
Mossotti, Ottaviano, 110
Mulder, Gerardus J, 122, 132
Müller, Hugo, 188
Müller, Johannes, 132, 136
murexide dyes, 27, 30

Nambu, Yoichiro, 225
natural theology, vii, ix, 60ff, 221
Ne'eman, Yuval, 222–3
neutrino, 219–20, 221, 223
neutron, 219–20

Newman, John, 18
Newton, Sir Isaac, 43–4, 84, 110, 184
Nier, A O, 216
nitrogen, transmutation, 25, 93–5, 126, 175

Oersted, Hans Christian, 110–11
oleaginous substances, 57, 122–3, 130, 132, 136
organic agents, 26, 28, 71–80, 127
oxalic acid, 160, 170

packing effect, 211, 213, 219
Paget, H M, 10
Paley, William, 60–1, 66–9
Pauli, Wolfgang, 219, 222, 224
Pelouze, Théophile J, 170, 173, 178
Penny, Frederick, 168–9
Pepys, William Haseldine, 35, 126
periodic law, 168, 179–80, 193, 195–6, 199, 201, 212, 221, 223
Perrin, Jean, 219
Philip, Alexander Wilson, 9, 76, 134
Phillips, Henry W, 10
Phillips, Richard, 134, 148, 153–4, 160; publicises Berzelius, 158; defends Thomson, 159; opposes Turner, 169; opposes Low's views on elements, 175
Phillips, Thomas, 8
Planck, Max, 218
polarity, 110–15, 123
Polkinghorne, John, 222
positive rays, 183, 204–7, 209, 217
Powell, Baden, 184
Powell, Cecil Frank, 220
Poynting, John H, 202–4
Prideaux, John, 163–4
primary matter *see* protyle
proton, 216–20, 222, 224
protyle, vii–viii, 45, 85–6, 90, 109, 162, 168, 170, 194, 198; Prout's usage, 37–8, 41, 105–7; word used by Crookes, 195; identified as positive rays, 205; connected with proton, 217
Prout, John, 2

Prout, William, x–xi, 146, 191; ancestry, 1–2; education, 2–5; search for medical practice, 5–6, 8; lectures on animal chemistry, 6–7, 23, 86–8, 95–6; made FRS, 6–8, 26; marriage and children, 8; Copley medal, 9, 18, 57; Gulstonian lectures, 9, 76, 118–20, 170; deafness, 9, 134; death, 10; portraits, 8, 10; analysis of organic compounds, 12–21, 57, 59; modifies slide-rule, 15; analysis of urine, 21–3, 25–32; views on diabetes, 25; designs hydrometer, 26; on urinary stone, 27; urinary diseases, 28–9; teeth and hair, 34; anatomical injection, 35; respiratory experiments, 35–6; speculative streak, 36, 109, 146; on sensations, 36–47; digestion, 47–59, 124–5; edits *Annals of Medicine*, 47; on blood, 47–9; chicken's egg, 47, 49–50; on gastric juice, 47; analyses of wines, 52; investigates food, 57; on cookery, 57–8, 125; coins term merorganisation, 58; on natural theology, 60–70; on meteorology, 64, 68–9; coins term 'convection', 64; religious views, 64; on use of iodine, 67; on living principle, 71; on potash, 96; on specific gravities, 97–102; announces protyle hypothesis, 106; reveals authorship of anonymous papers, 106; molecular theory of matter, 110–16; unity of forces, 114–15; adopts Avogadro's hypothesis, 116; debate with Henry, 117–18; comparison with Liebig, 122, 133–41; on arithmetical series, 123; primary and secondary assimilation, 125, 140; on milk, 125, 128, 140; gelatinification, 127, 137; theory of nutrition, 128; on chemical formulae, 132, 137–40; on oxalic acid, 160, 170; specific gravity of air, 165; works

Prout, William (contd)
   with Dalton, 166; protyle less than hydrogen, 169–70; final word on elements, 178; cited by J J Thomson, 194; in advertisement, 218; *writings*: *History of Physic*, 4; MD thesis, 5; *Inquiry into Nature of Gravel*, 28, 31, 52; *On Stomach Diseases*, 31–2, 127, 132, 135–6; *Bridgewater Treatise*, 32, 51, 56–7, 64–70, 76–7, 80, 109, 111, 117, 120, 131, 135–6, 170; *De Facultate Sentiendi*, 36, 41, 46, 59, 71, 85, 88, 96; unpublished book on digestion, 51
Prout's hypothesis, viii–ix, 36, 46, 105–6, 139, 158, 160, 174, 179, 193, 201, 217; Prout's early commitment, 85; Davy's influence, 89, 91; Thomson's influence, 92–3; distinction between integral and protyle hypotheses, 92, 150; Miers's influence, 94–5; Rainy's discussion, 154–5; Turner's opinion, 167–8; J J Thomson supports, 194; Crookes supports, 195; explanations of non-integral weights, 195, 199, 207; ideal law, 210–11; familiar to British children, 218; the Proutian atom, 219; in nuclear physics, 221–5
purpuric acid, 26–7

quarks, 223–5

radioactivity, 197–8, 207
Rainy, Henry, 154–8; view of Prout, 157
Ramsay, Sir William, 198
Redtenbacher, J, 173
reduction, 26, 124
respiration, 7, 35–6, 126, 138
Reynolds, J Emerson, 196
Richards, Theodore W, 201
Rigg, R, 175
Roberton, John, 78
Roget, Peter Mark, 8, 62–3
Röntgen, Wilhclm, 197
Roscoe, Sir Henry, 186, 188; advises Lockyer, 192
Ross, W H, 199
Royal College of Physicians, 5, 9–10, 76
Royal Society, 6–8, 18, 26, 47, 49, 52–3, 58, 61–2, 167, 189
Russell, William J, 188
Rutherford, Sir Ernest, 197–8, 200–1, 210–11, 212, 217, 224; model of atom 218; postulates the neutron, 219

saccharinous matter *see* carbohydrates
St Thomas's Hospital, 5–6, 56
Sakata, S, 220
Scheele, Karl W, 26
Schmidt, Carl, 14
Schuster, Arthur, 192
Schwann, Theodor, 56, 132
Scriabin, Alexander, 44
Scudamore, Sir Charles, 31
senses, the, 36–47
sight, 41–2
Silliman, Benjamin, 154
simplicity, ix–x, 93, 150, 221
smell, 41–2, 46–7
Smith, Edward, 36
Snow, Sir Charles Percy, 214
Soddy, Frederick, 179, 197–8, 199, 212, 217; concept of isotopes, 199–201; interpretation of neon as isotopic, 207
Somerville, Mary, 110
specific gravity, 97–102, 104–5, 143–4, 150, 154–8, 165, 167, 171
spectroscopy, 182–4, 186–90, 201, 214
Spencer, Herbert, 180, 195
Stas, Jean Servais, 106, 156, 168–70, 177, 191, 201, 210; weight of carbon, 172–3; rejects Prout, 174, 178–9
stereochemistry, 202
Stewart, Balfour, 192
Stokes, Sir George, 190

Strömholm, D, 199
sub-molecules, 119–20
sugars, 25–6, 58, 123–4
Svedberg, T, 199
synthesis, organic, 33, 74

taste, 41–2, 46–7
Taylor, Alfred, 51
Thenard, Louis Jacques, 13–14, 25
Thomson, George, P, 208, 212
Thomson, Joseph J, 182, 195, 197, 212, 217; cathode-ray research supports Prout, 194; work on conductivity of gases, 202; explains Crookes dark space, 203; work on positive rays, 204–5, 207, 209; scepticism towards isotopes, 211–12
Thomson, Robert Dundas, 56
Thomson, Thomas, 5, 7, 11, 21, 34, 47, 74, 83, 85, 92–7, 99–102, 104, 106–7, 119, 133, 143, 162, 168, 170, 173; supports integral weights, 144–5; attacked by Berzelius, 146–7, 158–60; attacked by Ure, 147–8, 153–4; *First Principles*, 148–52, 154, 157, 160–2; praises Prout, 149; his Pythagoreanism, 152; criticised by Rainy, 154–8; error with oxalic acid, 160; reputation, 161; criticised by Turner, 162–3
Tiedemann, Friedrich, 49, 52–3, 55–6
Tilden, William A, 202
Troughton, Edward, 16
Turner, Edward, 21, 154, 157, 160–1, 169–70; atomic weight determinations, 161–5; views on Prout's law, 167–8
Turner, Rev. John, 3
Tyndall, John, 64

Ure, Andrew, 20, 133, 147–8, 152, 155, 157; attacks Thomson, 147–8, 153–4
Urey, Harold C, 214–15
uric acid, 26–7
urine, 7, 13, 21–33, 59, 128–31, 135, 140

vapour density, 182, 193
Vauquelin, Nicolas L, 13
Villard, Paul, 197
vitalism, 20, 36, 70–80

Wakley, Thomas, 131–3
Walton, E T S, 216
Whewell, William, 63, 116, 183
whole number rule, 210, 212
Wien, Wilhelm, 204–5
Wilson, E D, 211
Wilson, George, 68
Witmer, E E, ix, 218
Wöhler, Friedrich, 23, 27, 74–5, 132
Wollaston, William Hyde, 8, 11, 15, 22, 26, 93, 97–8, 105, 133, 148
Wurtz, Charles, 182

Young, Charles A, 185–6, 192
Young, Thomas, 29
Yukawa, Hideki, ix, 220

Zweig, George, 223